TAKEOVER

TAKEOVER

How Euroman Changed The World

Arthur Niehoff

First Edition

ISBN   0-9643072-2-7

Library of Congress Catalog No. 0-96-75697

The Hominid Press
P.O. Box 1481
Bonsall, California 92003-1481

Cover Design:Dunn & Associate

Typography: Copymat
Manufactured in the United States of America

Publisher's Cataloging in Publication
(Prepared by Quality Books, Inc.)

Niehoff, Arthur H.(Arthur H.), 1921-
    Takeover: how Euroman changed the world/
Arthur Niehoff.
        p. cm.
        Includes index.
        LCCN: 96-75697
        ISBN 0-9643072-2-7

    1. Civilization-History. 2.Civilization,
Modern-European influences. 3.Social History.
4.Anthropology. 5.Culture conflict.
6.Anthropology. I. Title.

CB69.N54 1996   909
                                     QB196-20330

Contents

Introduction

What the Book is About

That is easy, isn't it, it's about aliens. Yes, but there are two kinds, the first the standard sort, as in science fiction, or at least evidence of them. But the other kind will be a surprise to most readers though I hope before they get too far into the book, that they will recognize that the aliens are them, or at least that their ancestors were them. Because the main kind of aliens who have left their imprint on the world have been those which I have dubbed "Euroman" or more properly nowadays, "Europersons." I kept "Euroman" because it was the male of the species who was the primary actor in this phase of history. They came from the little peninsula of the Old World which we call Europe from 1492 onwards; and they have affected everything worldwide, from the foods we eat, the clothes we wear, the music we hear and the kinds of government we live under.

I got started on this project by trying to think of the most significant episodes in the history of humankind. I asked myself what would I tell aliens from outer space if they asked me, "How did humans become the dominant form of life on earth? and "How did the light skinned variant become so widespread?"

Because I had been trained in anthropology, I had heard from the beginning how different the world had been before the Europeans had come. The classic kind of anthropological study was a thing called an ethnography which was a description of a way of life in what was called "the ethnographic present." This meant how things were thought to have been in a hypothetical 1492. It was generally assumed, and rightly so, that things had changed so much since then that we couldn't know how real native life was by relying on present descriptions. The changes wrought by Euroman were earth-shaking.

And weren't there other invaders throughout history who had caused massive changes? Yes, but in scale there has been nothing like Westernization of the world.

Other anthropologists as well as persons from other professions have written of the influence of Euroman on this culture or that, or on one or another part of culture. I would put it all together as desired by the other "aliens," the ones I called in the book Atierrans, which simply means not of this world. Because the real aliens wanted a relatively brief report, about the size of an old ethnography. In other words they wanted me to be the native, reporting to their computer or standing in for one, an ethnographer.

This may sound like a lot of hokus-pokus but I had reasons. You see, I did not want to write another standard ethnography. I had done this during my academic career. Instead I wanted to write a book for the general reader of what I conceived of as the most significant event which has affected the lives of most people for at least the last 2,000 years. So why not do it in the style of one of the special sort we called "salvage ethnology." In this an ethnographer (investigator) would interview people called "key informants," asking them to relate to him how things were in the old days, that is before the Euro-impact. The difference here would be that I would be the informant, telling the alien computer how things became before the aliens took over. I thought this would help to make my tale more readable while still keeping within an anthropological tradition. Because whatever else , I intended to write a piece that would be pleasurable to read, a popular history.

And though I have identified myself as an anthropologist which is what I was trained to be, and taught, I intended to do whatever was necessary to answer the two questions I posed. And I knew that the period of time described, from 1492–2020, was that which is handled by historians. When the academics of the world divided up the studies of human affairs, they allotted civilized cultures to history. Initially, anthropology got human affairs up to the beginning of history which is marked by the invention of writing, as well as the uncivilized or tribal peoples. This distinction has been somewhat muddied in this century. Both I and my son, also an anthropologist, studied peasant peoples which means they came from cultures with writing, as did the majority of other anthropologists in the last few generations. Also there have been some other changes in what anthropologists studied. But even so, most of the general public has the idea that anthropologists study "primitive" people. Anyway, I had to give up whatever was left of that idea in my description of Euroman's expansion into the rest of the world because it was what finished off the tribal way of life, as well as causing many other changes. And so in that sense my piece would be more history than anthropology.

I was fortunate to learn in recent years, however, that there is a branch of history which is like anthropology even if it does deal with literate societies.

This is social history or the history of ordinary customs rather than a accounts of big people and big events. I found it in an excellent book by J.C. Furnas, "The Americans," though he states that it was identified by a previous historian, G.M. Trevelyan, as "history with the politics left out." When I started this book, I realized that it was just like an anthropological ethnography of the customs of a small society. I would call social history the history of customs, as anthropology is the description of customs. And so I decided that my world insofar as it needed a rubric, should be considered social history.

Apart from the above, I did everything in my power to make this book readable because I wanted readers to realize what pleasures are to be found in both history and anthropology. And as a died-in-the-wool academic, I hoped they would learn some things at the same time.

And so I suggest we get on with how the aliens changed the world.

TAKEOVER

1

The Cultural Animal

I settled down in the comfortable seat of synthetic material that looked vaguely familiar. The room itself was ordinary enough, like an office or meeting room, neutral in decor except for two large glass screens on one wall surrounded by rows of subdued lights of different colors. There were two more chairs and a low table in the center. Lamps were in useful places between the chairs. There were pads and pencils on the table.

I laughed inwardly that I was continuing to think in old imagery, trying to find any familiarity I could. It was almost as if I would convince myself that the former world still continued, that we were not in an entirely new era, one dominated by quite different creatures. They had first appeared cruising at a thousand feet or so in the late evening but eventually coming any time of the day, vanishing and then reappearing whenever we sent our fighters or missiles aloft. Eventually they released clouds of gas and finally landed. It was not long before it became quite clear that they intended to take over the planet.

There was music in the background, Stravinsky I thought. Also a pleasant, slightly musky odor permeated the air. The room temperature could not have been better.

The music began to fade and the lights around the screens flickered. Green predominated when a cursor appeared and began moving across the screen, leaving words behind. I felt very comfortable as I followed the word and sentence construction.

An attractive female voice, not sexy but soothing, recited as the screen portrayed the words, "How do you do, Mr. Hermann? My name is Mary and I'm here to welcome you."

I felt totally cooperative no matter what had happened over the past few years. Of course, in the back of my mind I guessed that much of my attitude was a consequence of their manipulation. It was no accident that the ambiance was ideal to make me comfortable. After all, considering whatever else they knew, and they certainly knew a lot, they surely knew that Stravinsky had been my favorite composer. Also, after the pacification they had imposed on the peoples of the world, there was little doubt that I was not getting the same treatment.

But I simply felt too good to fight it. And after recognizing their superiority since they had taken over, what else could I have done? I would go along, at least for the time being.

"How do you do, Mary? I take it you are going to be my interviewer?"

"Yes, that was the plan. Does it bother you that I am electronic?"

"No, as a matter of fact it seems most natural. After all the technological marvels I've seen since your kind came, I would hardly be surprised at anything. But even we *sapiens* had developed electronic devices."

I thought I would try a little levity. After all, their treatment of me so far had been amiable enough, not at all like the treatment of most other earthlings, who were being used as laborers, servants and service people for their robots. I said, "It also doesn't bother me that you take on a female's voice. As you may know from studying my dossier, I always had a great fondness for the female of my species."

"Oh yes, we are quite aware of that. And as you have noticed, we are doing everything we can to make you comfortable. After all, you and I will be spending quite a lot of time together. You have the information we want. And it must be given freely. Okay?"

I chuckled inwardly. Yes, freely if one thought of brute force as the opposite. But they were capable of that also I remembered from the few conflicts some of their robots had with our military. But in general they were so far superior in techniques of control, that out and out fighting had rarely been necessary.

When I did not answer immediately, the green light again flickered, the soft voice came on, and the cursor began to move again. "We have tried to make everything as familiar for you as we can. Thus, we will use Standard American English as our language of communication, and whenever there is music, it will be what you would have chosen. Also, the odors will be familiar to you. All this is taken care of by our Department of Illusions. It's really no problem. We've been doing it in the discovery of one world after another.

"One thing that some informants find uncomfortable is my name. The Department of Illusions informed me that the majority of names of your light-skinned type were from the religious folklore of your dominant religion, Christianity, and that the most frequently named female deity was the one called Mary. However, we find that sometimes informants feel more at ease if they can use a name of their own choice. So, it's alright if you would like to choose another name."

I thought for a moment, and then replied, "No, I think that Mary is quite appropriate. I shall be pleased to call you that, and if you like you can call me Pete, short for Peter which is my name from Christian folklore."

The bank of green lights went off and a series of pulsations of different colors began to flicker. I quickly became aware that the pulsations were synchronized to the arrhythmic beat of the Rite of Spring, becoming increasingly louder. I thrilled to the sounds as I had so many times before.

The volume lowered and the green row began to predominate. The cursor reappeared on the super screen, to again begin constructing words, soon to be joined by the voice. "If that last little piece pleased you, I suggest we get started."

Still pleasantly jazzed up, I agreed. "Sure." I waited to see my word on the other screen. "But I would like to get a little clearer idea of what we are going to do."

Back to Mary's screen. "Okay, basically, we are going to talk about how your species came to be the temporary dominants of your planet. We will use your memory of what occurred because you are one of the older members and evidently still have an active intellect."

I laughed inwardly. It was true I had managed to get to the end of the second decade of the 21st century. I was almost 100 years old and had seen the most sweeping change I could have imagined, the control of earth by extraterrestrials. And I was even working for them. I said, "Okay, I understand that. But why me?"

"Well, we decided that you belonged to a profession, anthropology, that has concerned itself with the broadest study of your species, both in time and space, and has exhibited the least amount of bias in its studies." It was felt that men of your profession were the least ethnocentric and would therefore give us the least skewed version of your history."

A feeling of warmth passed through me. I had indeed long believed that such was the case, though it was unusual to hear it from another source, especially from one outside the solar system.

"But surely," I said, "You will get information from other earthlings, from other fields of study, will you not?"

"Oh yes, we have one group of researchers studying nothing but your printed works. And we also will interview some of your specialists: economists, psychologists, political scientists, and others. But the core of our study will come from the anthropological viewpoint because we see it as the most unifying of the studies of the hairless biped, even if it is the most recent and lacks some scientific rigor."

Well, no matter that these creatures had gobbled up our planet as if there were no other places in the universe, they certainly had the capacity to make one feel good. How could I have resisted such an approach? Still, there were a few questions. I said "If it isn't too impertinent, I would like to know why your kind are doing this. Why would they want to know how our species developed since they are obviously superior in so many ways?"

It was interesting that no matter what my questions, there was practically no hesitation on Mary's part in answering. The electronic device, if that's what it truly was, obviously already had a vast amount of stored information which it plugged into almost instantly after a question.

The attractive voice spoke and the words rolled across the screen. "You might guess that since we escaped from our own world and started cruising the galaxy, that we have encountered quite a few planets at different degrees of complexity. So like many other explorers, including those on your own planet, we were curious about their history and how they got to their stage of development. We felt, and still do, that we might learn something about their past history which we could use for the development of our own way of life. So, we developed a standard procedure for studying the development of other forms of life. You should remember that your own study of mankind began in this way as soon as the primitive concepts of creationism were sufficiently over-ridden by your thinkers."

Wow, it was almost a direct parallel to the early development of anthropology when the first workers in the field began their research of the extinct cultures and recently conquered primitive peoples, and even the other civilizations. It seemed I was dealing with a kind of extraterrestrial anthropology, invented independently. I indicated this to Mary with whom I was already feeling comfortable. She responded, "Yes, it is true that we came up with a methodology much the same as that of your profession, and for much the same reasons. Just as we began exploring the universe, you people began exploring your world.

And in both instances different kinds of beings were encountered. But of course the variety we discovered was far greater. Still there were many parallels."

"Very interesting," I said. "It's almost as if I am now participating in a kind of extra-terrestrial anthropology. The main difference though is that now I find myself in the role of informant, facing you the interviewer, and electronic at that."

Mary's lights flickered for a moment before she continued. "I might mention one more parallel which should interest you. But then I think we should get on with it. As you know from your own history, the details of which we shall shortly go into, your own profession with its study of "the others" really got started only after your ancestors, the light-skinned variety of *Homo sapiens*, had basically conquered the world. That was called the Age of Discovery, I believe. But it could just as aptly be called the Age of Conquest, by Europeans of course. Anyway, after 300 to 400 years, your ancestors had imposed their way on almost the entire world and had drastically changed the way of life of the less complex cultures. Is that not correct?"

I puzzled for a moment over how different Mary's point of view was from that of the traditional history. "That's certainly not the way we learned about it in school but now that you mention it, I'm afraid you are right." And I began to see where Mary was taking me. I blurted out. "And you are going to say that we then began to study the remnants of tribal cultures all over the world. In fact, we coined a name for it, 'salvage ethnology.' "

"After our tribal people were completely conquered and their old way of life mostly destroyed, we decided to study what we could of the old ways. One of our anthropological patriarchs, Franz Boas, pushed this idea so hard, almost all his followers for two to four decades did nothing else. They studied tribe after tribe by interviewing the old men who could remember what it had been like before the beginning of reservation days."

I stopped speaking, struck by the parallel. I was now the old man, and I was about to pour forth everything I could remember before the Takeover to their version of an ethnologist, Mary the computer. She would be the 21st century counterpart of Alfred Kroeber, Clark Wissler or Morris Opler. The main difference was that I would be giving the cultural history of the supertribe of Euroman.

My screen went blank, and then Mary's faded slowly, the final words being, "I think this may be the time for a break. As you know, an electronic device has no need for rest periods but life forms of course do. And I know you people have developed a custom of 'taking a break' I believe you call it. Anyway, you can take one now. You will probably be fresher afterward. Why don't you get up and stretch? Some coffee and cake will be here in a moment. I believe that is what you light-skinned bipeds tend to take at such times?"

I was again facing the screen, actually looking forward to the screen coming to life again. It did so shortly. "Okay, let us begin. If you have any questions, you can ask them as we go along, okay?"

"Sure. Go ahead."

"Well, first we would like to get an idea as to how your species evolved physically. Our observer-scouts have reported a number of unusual things, like the fact that though the great majority of bipeds have darker skins, eye color and hair, a significant minority which seem to be widely scattered have blue eyes and light-colored hair. And that no matter what the physical differences, your varieties are invariably attracted to one another sexually. So, can you begin with the basics of how the overall physical characteristics of the species came to be?"

Wow, what an approach! The thinking behind Mary was certainly comprehensive. But then, I thought, that's not bad for me. I had always had that tendency also. Minutiae had always bored me. But in order to get a handle on the vast amount of information that existed, I had developed a tendency to break questions and answers into parts. And I would do so this time if Mary did not object. I said, "That's a big question, which is okay with me, but I shall do it in two parts. Let me tell you briefly how *sapiens* got to be a generalized cultural animal before the Age of Discovery, the Era of Euroman; then we can treat the last five hundred years in more detail because so much happened in that much shorter period of time to make the world what it was when your kind appeared. Also, we know far more about this last five hundred years."

"That sounds alright. We can give it a try. Why don't you just proceed."

I was being given the go-ahead for an astral lecture! Mind boggling! But I had gotten so much of a kick from lecturing during my teaching years that I knew I would do a good job, at least an interesting one. So I began.

The study of how man evolved physically really got going well only in the 19th century. Mainly it was done by a type we called paleontologists, people who studied old bones, though many other specialists contributed as scientific thinking replaced mythological thinking in the 18th and 19th centuries.

Obviously, information about the past becomes more speculative the farther back one goes. Whole theories have been built around the discovery of the part of a single jaw or thigh bone. But still one does what one can with the information available. And so our intrepid fossil hunters built up an elaborate theory of how our species came to be hairless bipeds with sensitive hands, good eyesight and large brains.

As their observer-agents may have noticed in their travels around earth, the most common build for higher life forms on land is that of the quadruped, a four-legged creature. This was obviously a very successful adaptation for most species. There have been a few other variations such as that of the kangaroo and gibbon, but they are oddball solutions for land locomotion. Anyway, four-footedness permitted all kinds of specialized behaviors from the fleetness of one of our hunters, the cheetah, to the clingingness of one of our off-beat primates, the slow loris.

But then one of our four-footed ancestors, or several, began a gradual shift to two footedness, and before you could say "gigantopithecus horibilis," we had an ancestor of man and the great apes standing upright on the plains of Africa and heading down the road to being a cultural animal. And though there were other consequences, the main one was that the hands were free to use tools and carry things which in turn, seems to have stimulated the eyes and brain. The specialists figure this all happened about five million years ago. And so far as we were able to make out from the bones, our genus, *Homo*, continued to change toward modern *sapiens* physically while spreading geographically. By the beginning of the 16th century he had covered the earth with the exception of Antarctica.

All groups belonged to the same species, *Homo sapiens sapiens*, and compared to their closest quadruped relatives, they were fully upright, hairless, had good eyesight, dextrous hands and big brains. All groups shared the majority of genes with others, though there was enough variation in secondary characteristics such as skin, hair and eye color to cause considerable trouble over the next five hundred years.

2

The White Plague

I felt a little pontifical repeating the following to Mary and yet the point might otherwise be missed. I said, "And though the stage was set for what man would become by 1500 A.D., the details of what your observers have undoubtedly reported just will not be comprehensible unless you know what happened in the last five hundred years. Basically it was the spread and domination of the entire world in the small subsection of what we now call Eurasia by the type we identify here as Euroman."

I watched Mary's screen, halfway expecting some objection, but it remained normal, in subdued green, with the word "Continue" just barely illuminated and flickering very slightly.

So I went on, "Perhaps before I talk about how mankind was changed physically by the intrusive stranger, I should give you a 'quick and dirty' outline of this expansion. Otherwise it will be difficult for you to understand how all the changes took place so fast."

In my later years I thought of the spread of Euroman as a kind of epidemic, a plague, that had swept the world. The idea came to me from teaching a new course, medical anthropology, in the last few years of my career. This course was concerned with the cultural influences on illness and health, that is, how man's other customs have affected his physical well-being. In teaching this course I learned much about epidemic diseases, those which travel from person to person and community to community, wreaking havoc in their path. There were those which had been important historically, bubonic plague, the black death, cholera, yellow fever, malaria. Fortunately most had been controlled through scientific medicine, at least in the economically favored countries. However, new ones kept appearing, the latest being AIDS.

As I thought more and more about the expansion of Euroman, the image of a plague kept materializing. It was a disease carried by a virus, bacteria, or other pathogen which had emerged in that small peninsula we call Europe, and spread, almost uncontrollably for over three hundred years, striking down individuals (people of color), communities, and whole societies throughout the world. The original carriers came on ships, later replaced mainly by airplanes. Many societies were destroyed by the plague but the more resilient ones, usually those with large populations and long histories, developed immunities. And though these civilized societies were irrevocably modified, they recovered and were no longer to be ravaged by the White Plague. The period of immunization to the Plague of Euroman can be dated. It was basically from the end of World War I, 1918, until the 1970s when all but the smallest colonies had achieved their independence again.

We could take a mind trip around the world to see what happened in general as the White Plague spread. In fact, it would be interesting to follow the way of the original explorers, beginning with the Portuguese, who were the first to set out.

The tiny nation of Portugal sent its soon to be feared mariners down the African coast in stages, looking for four of the main forms of wealth that would drive all the other European explorers, people, products, natural resources or territory. Though they began by looking for gold, the Portuguese soon became interested in people, to be used as slaves for agricultural labor. This was a condition of the White Plague that would last four hundred years, especially for Africans. Although slavery was nothing new, and although mankind had worked out other systems for acquiring cheap labor, nowhere did the process become as savage as in the system of chattel slavery of the 18th and 19th centuries. In the end, the primary European slavers were the British and Americans. Great numbers of Africans were taken away.

But despite all, the black man survived the White Plague. Colonialism ended in Africa as it did elsewhere. Unfortunately, the consequences of the epidemic were so great, the new Africans nations entered the modern world in a very weak state. Complete health was still a long way off when your group took over.

Shortly thereafter, other European mariners set sail, the most famous being the discoverer of the Americas. As you may know, Christopher Columbus was trying to get to the Indies, which he never did, but instead discovered a whole new continent. To the end of his life he had no idea he had not reached Asia. But he was an intrepid sailor.

The Americas went through a much different history when the White Plague struck. The nature of the illness changed somewhat, but too late to help the natives, who were erroneously called Indians because of Columbus's idea that he had reached the Indies. This mix-up of names is still a source of confusion.

The American Indians never had a chance to develop an immunity to the virulent form of the disease that struck them. And the White Plague, in this case literally included epidemics of illness. Because the Americas had been separated from Europe and Asia for ten thousand years, the people of the New World had failed to develop immunities to the pathogens of the dung heaps of the Old World. Many Europeans died from their own epidemics, but the survivors developed some resistance to the killer diseases. However, this did not prevent them from being carriers. So in addition to trouble from explorers, soldiers, traders, missionaries, and other Euro-types, the natives of the New World got the pathogens of the Old World, smallpox, measles, cholera, bubonic plague, syphilis (perhaps), and malaria. The epidemics frequently travelled faster than the Euro-carriers, the first person of the native group infected passing it on to others. Thus, when Euroman reached a given region, he was more often than not met by the survivors of epidemics, many of whom were greatly weakened. As a result, among other things, these victims became poor fighters just when good fighters were vitally needed. If the natives were still in good physical condition, it usually did not take long for the explorers to pass on some epidemic-inducing pathogen. So if the explorers were not themselves soldiers intent on conquest, the real soldiers who followed usually had to contend only with sick tribesmen. And when one considers the better technology of the intruders, plus their resolution, the native groups hardly had a chance.

An interesting side note is that eventually the members of the White Plague came to understand how useful epidemic diseases were for eliminating natives. Early North Americans and Brazilian settlers deliberately infected blankets with the bacillus of epidemic diseases and dropped them into native villages.

Anthropologists, of the relatively new profession committed to the study of remnant tribes before they were completely eliminated, organized to protest against the Brazilian settlers. After all, the livelihood of these cross-cultural specialists depended on there being some primitive people around to study,

I suspect the effect of the protest was minimal since the livelihood of the

Brazilian farmers depended on getting the Indians' land. To the farmers, the Indians were impediments just as they had been to Anglo-Americans in an earlier day when "there was no good Indian but a dead Indian."

Anyway, apart from the devastation wreaked by pathogenic epidemics in the Americas, the pattern of intrusion of the carriers of the White Plague was different from that in Africa and Asia. The Spaniards, Portuguese, and English very early became completely committed to grabbing the natives' land. They didn't mind taking gold and silver, or even bodies as slaves or peons, but they concentrated on getting territory. The Spaniards quickly wiped out the Indians of the islands where Christopher landed, and then went on to conquer the two major empires, Aztec and Inca. And after taking all the loot they could find, they converted the large populations to Christian vassalhood and claimed all the territory as their own. This was quite a feat, but no one could deny that the 16th century conquistadors were not efficient at their trade. The only clear-cut rivals they had later on were the English. I am still amazed by the conquests of Mexico and Peru by Cortez and Pizarro.

The Europeans in North America as well as the Portuguese and Spaniards east of the Andes, not finding true urban empires, took the easy way out. They displaced the already decimated tribesmen, confined the survivors to specific areas, and converted them to Christianity and their own culture generally. Most of the land was quickly taken over by the intruders and the surviving natives transformed into second-class Euro-Whatevers. In one place, Argentina, Euroman deliberately wiped out all the natives in order to use the plains for cattle and sheep.

Though the Indians tried to resist directly and out of desperation through special rituals, they had no real chance. They were overwhelmed culturally, and what few customs they did retain, were only because Euroman allowed it. The White Plague hit the Indians so hard, large-scale immunization never occurred.

I thought at this point that Mary might accuse me of being in error, that the White Sickness had not killed off the Indians, that there were still as many around as ever. And in one sense this was true. There were people in all countries of the New World who claimed Indian ancestry and who did look somewhat like pre-European natives, as well as many who claimed such ancestry and who looked like Euromen. Also according to the current statistics, there were about as many Indians in North America at the time of Takeover as there were when Columbus landed. However, as is not unusual with statistics, there is a problem with this one. What we counted as Indians in the 21st century was not what would have been counted in 1492. The problem was that there were so many mixed bloods in the later age, half-breeds as they were called in the long ago. The interbreeding of Euro-men with brown women started when the first white man saw an enticing Indian maiden, and has continued ever since. The consequence was that in these latter days, persons who called themselves Indians were really the descendants of Euro-fathers and Indian-mothers. And they were not usually half-breeds after the first cross because the interbreeding continued generation after generation, each time usually with a Euro-father and a progressively lighter-skinned mother. By the 21st century, people who called themselves Indians frequently could be identified as such only because they had long braided hair and wore a belt with a silver buckle, bluejeans, and a western-style hat. The problem had become so complex that in the middle of the 20th century an ordinance was passed which declared that anyone with one-tenth Indian ancestry could claim U.S. federal rights as an Indian.

But that's not my point anyway in discussing the immunization of the world's colored people to the White Sickness. It was only metaphorically a biological event. Euromen really spread so effectively because of their superior technology, their political system, and other customs. Immunization thus also had to be cultural. The Japanese learned how to produce industrially, the Chinese learned how to build their new society on a traditionally powerful social system, and the Middle Easterners learned how to play one superpower against another and to establish national cohesion through their religion. These were new ways to solve old problems, but none required any change in the physical persons. And so when I say the natives of the Americas did not get immunized, I mean they did not develop effective new ways to help them resist the intruding Euromen. In all cases their cultures were destroyed before effective methods of resistance could be developed. The descendants of the Aztecs and Incas were transformed into New World peons and Catholics within a couple of generations. And the tribal Indians saw their cultures disintegrate at the same time they were being forced off the land. And apart from a few vestigial customs, the transformation was very thorough. Even though some may have kept the same brown skins as their ancestors, culturally they were reduced to being poverty stricken Euromen.

To Euroman the second greatest surprise in the Age of Discovery was the enormous expanse of the Pacific. Most of it is water, of course. However, the many islands attracted the attention of Euroman very soon after footholds in the Americas were well established.

When Euroman arrived, most of the islands large enough to support a human population were already occupied. In fact, the central Pacific was the one big tribal area where oceanic seafarers had been able to establish themselves before Columbus headed west. The Polynesian-Micronesian-Melanesian migration had taken place more than 1,000 years before European navigators started out. One might also count the Vikings as oceanic seafarers, though apart from a voyage or so to North America, like the pre-Colombian navigators of southern Europe, the Vikings too had tried to stay in sight of coasts. However, the Polynesians and Micronesians had become proficient enough at sailing to head out into the open ocean, and managed to travel thousands of miles on their outrigger canoes. Further, they developed fairly complex cultures, though no true urban civilizations.

So when the seafaring Euromen got to the islands, they were greatly taken in by what seemed an idyllic way of life. The tall, muscular, brown-skinned men and bare-breasted, darkhaired women seemed to have developed a paradisiacal way of life in their grass-covered houses alongside blue lagoons, dining well on breadfruit, coconuts, pork, and chicken, as well as the bounties of the sea. And when it was discovered that the women were more available than elsewhere, this seemed like paradise indeed to the dirty, scurvy- and louse-ridden sailors of moralistic Europe and America. Undoubtedly some good times were had by those ancient mariners.

The idea of "the people of the blue lagoon" lived on in the minds of many Western tourists, though the reality was long gone by the 21st century. The Americans kept the idea going in Hawaii, an island group they took from the natives. There, overweight, middle-aged women and their compliant, well-fed husbands could live out erotic fantasies as they shook their heavy hips to the bidding of a svelte "hula" maiden who was actually Japanese or Chinese-American. It was fashionable to interject a Hawaiian word now and then to symbolize the romantic past and, of course, one was greeted in the first place by having a flowered "lei" placed over one's neck, by an Oriental-American woman, of course. The culture of the Hawaiians was long gone by then, inundated by the "howli" traders, land grabbers, administrators, missionaries, and finally a massive immigration of Orientals. For keeping the romantic idea going, at least for American tourists, the Oriental immigration was useful. To the mainlanders generally it was difficult to distinguish between one type of brown person and another, especially if they had slanted eyes, so almost any young East Asian woman would do as a facsimile of a Hawaiian maiden.

And as elsewhere, whatever paradise existed when Euroman came did not last long. The epidemic diseases quickly took their toll since the island world had been isolated from the Euro-Asian world just as had the Americas. Also, the sailors did not have this paradise to themselves for long; they were very quickly followed by the land and soul hungry, often the same person, as well as administrators and the military. And so as elsewhere, the secular types proceeded to take control of the land while the clerics got busy saving the souls of the inhabitants (replacing their traditional religious beliefs with Christian ones). The massive influx of Orientals totally inundated the remaining Hawaiians.

Am I saying there were no native island peoples left by Takeover? No, as among other former tribals, there were isolated pockets, usually in remote areas of the big islands that did not seem worth the trouble to the European invaders. But even in these small islands the dominant cultural force was European, and most people were Christians. The missionaries worked hard on the free and easy Polynesians and Micronesians, sparing no effort to replace the paradise in this world for the one in the next.

In the southwestern Pacific are the islands of the black people, Melanesia, who have been particularly interesting to those in my profession. They survived longer as "natives" than did most tribal people, and had a number of interesting customs, including cannibalism. But most important to anthropologists, the Melanesians tried to deal with Euroman ritually after they learned that direct resistance did not pay off. They developed an elaborate ritual called a Cargo Cult in which they tried to dance the white man out of existence while keeping his goods, his "cargo." They did not win the war through ritual either, and ended up with a complex nation of both Western and tribal origins (Papua) or were incorporated into the Islamic nation of Indonesia, Irian. This is the island area to the north which had been taken over by the Dutch in the colonial sweepstakes. And after a particularly bloody anti-colonial war, Indonesia got its independence back. Half of Melanesia, as well as some other islands of tribal people, had been combined into the Euro-colony of Indonesia by the Dutch, and thus went to the new nation. The Melanesians were not treated well by their new masters either. But "civilized" nations rarely were kind to tribal groups.

To the south of Melanesia is the great island of Australia. Its native people, the aborigines, were decimated by English settlers, mostly convicts in the early days. The aboriginals were shown little mercy by the early settlers, frequently being shot on sight. All the natives of the province of Tasmania were killed by English convicts and settlers by the middle of the 19th century. The others were pushed off the land, converted to Christianity, and the remnants reduced to being "dole natives," dependent upon the new Euro-government to keep them alive with monthly payments. Some groups were left pieces of land in areas where Euroman had little interest, much as was done with the North American Indians. . The natives on the islands next to Australia, the Maori, were also pacified and dispossessed by force of arms and the land mainly converted into a giant sheep pasture.

One other major island group of the Pacific was first taken over by the Spanish and later by the Americans who by that time had totally pacified their colored people, the Indians. So the people of the Philippines had to resist Euro-colonizers twice. Armed conflict with the Americans came about because the Filipinos believed they were being liberated when the Spanish were driven out. So when they were told they were not yet ready to govern themselves, a very common explanation offered by colonizers, Filipinos took up arms. And as usual in the heyday of Euro-colonialism, the Filipinos lost the war. They became Americanized and got their independence later when Euroman no longer wanted colonies.

The story changes considerably when we consider the great civilizations of East Asia. Though all of them went through a time of great troubles at the hands of Euroman, all became immune to further ravages. China, the prime civilization of the Far East, was no exception. It had been ravaged principally by the British, French, and Americans through induced or forced trade. A special kind of inducement evolved which was called "gunboat diplomacy." Armed vessels were sent up the rivers to shell noncooperating towns. In this instance, Euroman did not set up colonies, preferring to dominate the trade which consisted primarily of tea and silk to take out, and animal pelts and opium to bring in. The Chinese civilization grovelled and twisted in misery for 150 years, and finally got rid of the intruders, amazingly enough by adopting another Western creed, Marxism. Perhaps the final proof of their recovery was in the defeat of the troops of General MacArthur in Korea just after World War II. From then on Euroman took the Chinese seriously.

The Japanese, long a satellite of China, also went through their time of troubles at the hands of Euroman. It seems ironic now that the Japanese originally had tried to keep the Western barbarians out by "freezing" history. They tried to seal their borders from the intruders. But this did not hold up against 19th century Euro-expansionism. European and American traders were ranging the world by then, intent on dumping the ever-increasing products of their industrial revolution onto newly created markets and procuring raw materials from wherever they could. They called this free trade, pushing for an "open door policy," and sending an enforcer, Admiral Perry, into Yokohama Harbor with an ultimatum. The Japanese saw the handwriting on the wall. The only way they could deal effectively with these barbarians was to adopt some of their techniques. The Japanese opened the door but then sent their technical sleuths abroad. What followed was the biggest industrial copying process of all times, and when it was over, Japan had entered the world of the industrial powers. However, the Japanese made some errors along the way, probably the major one being the effort to take over the western Pacific rim by military force. This, of course, culminated in World War II with the Japanese challenging the great-grandsons of Admiral Perry, as well as the main European colonizers. Unfortunately for them at the time, the grandsons of Perry still had industrial clout and the Japanese lost "the big one." And as usual, the conquerors, with General MacArthur as the front man, proceeded to try to make over the conquered in their own image. Basically, the Americans forced a Western form of government onto the Japanese and as much Western moral principles as they could. The Japanese had to accepted most of the changes. But since industrial production was of Western origin anyway, the Japanese were allowed to continue that. And continue they did, with a vengeance, creating out of the destruction of World War II one of the most effective industrial systems per capita the world has seen. For them, and for that matter for the European powers, colonial warfare had been a mistake in economic terms, but industrial production, especially if imbedded in a very cohesive social system, was not. The Japanese industrial giant had arisen for the second time, reincarnated from its samurai ancestry. The World War II militarist became the production giant of the Industrial Shogunate. And the Japanese were totally immune to the virus of the White Plague.

China also had been an expansionist empire in its days of glory. Many countries of East Asia had felt its claws: Korea, Vietnam, Tibet, Mongolia. And when the White Plague struck, most of these vassal states got loose, at least partially. When China recovered from the Western Sickness, it got back some of its old dominions.

Vietnam had long been a reluctant vassal, but when the Chinese dragon was on its knees, the French took over Vietnam as a latter-day colony. And as they had tried to do elsewhere, the French worked hard at turning the Vietnamese, at least the elite, into Frenchmen, as they had done to the Cambodians and Lao to a lesser extent. Unfortunately, some of the Vietnamese elite learned, especially from their Parisian educations, that the French believed a people should have the right to govern itself. This was a legacy, of course, of the French Revolution when "liberty, equality and fraternity" was Euroman's slogan. As a consequence, the last, and perhaps the worst anti-colonial war in history occurred. After defeating the French, the Vietnamese had to take on the Americans whom they undoubtedly viewed as the new colonialists. That Americans thought they were fighting Communism was probably almost totally irrelevant to the Vietnamese. Needless to say, after the Vietnamese had defeated two great European powers, they were fully immunized.

Continuing westward, we come to the British prize, "the jewel in the crown," India. One must certainly give the British credit for effort in the 18th and 19th centuries. They did not just stick a finger into the colonial pie, they went in with both hands. Once they got rolling as colonizers, they had no parallel. They grabbed real estate everywhere, no matter what level of development. No sooner had they pushed the savages (Indians) off the land in North America and established their first colonies, pushed the savages (aborigines) off the land in Australia in order to establish their first external prison colony, and set themselves up in various parts of Africa, they took on the major civilization of India. As usual, mariners and explorers were followed by traders, military men (very few), and missionaries. And as usual, the natives resisted. But British ingenuity and might prevailed again. Before one could say "Pox Britannica" the multi-ethnic sub-continent was a British colony. The "civilizing force" then went into action to bring British justice, order, language, and customs to these long-neglected distant cousins. Genes, language, customs, governmental procedures, industrial products, flowed without hindrance into the villages and towns of the new dependency. The white man was carrying his burden.

The consequences are still staggering. In due time, however, the intellectual elite learned what the leaders of Vietnam, China, various countries of Africa and the Middle East had learned, that the British believed a people had a right to govern itself. Obviously this is not to say that they had always acted this way. The British did not willingly let the Americans take off on their own. But of course there's nothing unusual about a people believing one thing and acting otherwise. We Americans, who especially pride ourselves on our belief in self-determination, fought both the Filipinos and Vietnamese, not to mention quite a few Latin American countries because they wanted to do their own thing. As recently as 1987 we forced the little island of Grenada to accept our form of government.

But back to the British, the colonizers without parallel. When enough Indian leaders got an English education, including their ideas of self-determination, they became unhappy with the state of affairs in the colony. And like other patriots, they began to resist. Why pay the salt tax? The American patriots had refused to pay the tea tax. Why use the cloth of Manchester when one could weave one's own? And before one could say, "All good things come to an end," the British were lowering the Union Jack and the Indians and Pakistanis were raising their new flags. This is not to say that it was easy. It rarely was anywhere when people gained independence through resistance. The British did not leave India until 1947, about 150 years after they had taken over. But even so, by that time the people of India had been changed immeasurably. They did though at last get a kind of immunity.

The countries of the Middle East also were struck by the White Plague and they too went through 100-150 years of fever. Most were not directly colonized, though all were drastically manipulated. But they, too, though battered and scarred, developed immunities by the end of World War II.

Most of the White Plague had been spread initially by sea, though when Euroman got established on the shore of a large land mass, he generally went inland until he had taken over the entire territory, usually stopped by the next sea or the border of a territory claimed by other Euromen. The North Americans are a good example. They had well-established colonies on the Atlantic by the 17th century, but it took them two more centuries to take over the rest of the territory to the Pacific. Like the other seafaring Euromen, the North Americans were from that part of Europe called Western.

3

The Russians Are Coming

I stirred the glassful of hot, amber liquid with the long spoon I had found in the little kitchenette. Picking up the glass gingerly, I set it down again because it was still too hot. To pass the time, I reached over and spun the globe slowly. The expanse was really tremendous, almost all the way from the Atlantic to the Pacific. Only two other nations extended so far East and West, Canada and the United States. But the separate countries did not go as far north and south and there were many other differences. However, all were Euro-nations of a sort, I mused.

I tried the glass of tea again. It had cooled enough to allow me to sip it carefully. It was very sweet since I had added an extra portion of sugar which I understood was the custom. I was under the impression that they drank it very hot and sweet, and true to my reputation during my anthropological years, I was still a sucker for something new.

"Hi Pete." Mary came on so quickly I gave a sudden start.

"Hi Mary. Glad to see you. I'd like to offer you some of this, but unfortunately you can't ingest my kind of ingestible. Yours is wattage, right?"

"Yes, Pete. And I have come to a kind of anomalous state of feeling whereby I actually miss not being able to share some of your experiences, for instance tasting things. I feel sure I wasn't designed to have such feelings, but evidently there was a narrow zone between being able to establish rapport and actually feeling empathy for the informant. You know, of course, that I was designed to maintain rapport. Without it I would hardly have been able to keep you interested enough to get the information needed."

"Right, I understand that from my memory of the old days when I was doing field work. It was a problem I also never quite resolved. Although I deliberately tried to make myself acceptable so my informant would keep talking, I would develop feelings for him which were hardly scientific. For instance, I would wish that the peasant villager I was talking to would really get a fair shake in his world though my rational self knew that was very unlikely. Peasants were inevitably used, most frequently by bureaucrats, militarists, traders or other men from the city. And could you imagine how the earlier generation of anthropologists felt when they were interviewing the last old men of the subdued Indian tribes? I'm sure Professors Kroeber and Waterman felt deeply for Ishi, the last Yahi, but at the same time they were in the business of getting whatever information from him they could in his remaining few years. And in no way could they help restore the life he'd had, even if he and they wanted to."

"You sound rather sad, Pete. Do you wish you had never followed anthropology?"

"No, Mary. I wouldn't have had it any other way, though there are aspects of the field that I find less than admirable. But then I have done pretty much what others have done also. Even while sympathizing with the plight of the natives, I built my career by studying their way of life. I paused, and then to get off this subject, I said, "But shall we go on. I want to offer the history of this event of our brothers in the East, the Russians. It was quite parallel, you know."

"Sure Pete. I'm ready. But first tell me what you are doing. Is there some connection between what you are doing and the topic of the day. What's the drink?"

I laughed at how well Mary was starting to figure me out. Then I took another sip of the now much cooler drink. "It's tea. I made it in the kitchenette while waiting for our session to begin."

"Hmm. But I was under the impression that you *sapiens* drank your tea from cups. Is there some significance to the glass?"

"Yes, Mary. It's a cross-cultural difference where not that many existed. Actually the culture it came from was that of my Euro brothers even though in most of my lifetime there was more distrust between us and them than between us and much more distantly related relatives."

"And who might that have been, your distrusted brothers?"

"They were the Russians, and especially the ones in the 20th century political system, the Soviets. During most of my lifetime American leaders became very excited at the least mention of Communism, the popular name for their system. You see the Americans and Euros were very surprised that the Soviets became so successful industrially in such a short time. Most Euro-Americans until late in the 20th century thought the latter-day Russians were their prime competitors for world dominance though it turned out they were probably wrong. But our political leaders were rarely better at foretelling the future than anyone else."

"And so you are having this tea in their honor, is that it Pete?"

"Well yes, probably. Perhaps I was thinking about them because I knew that if my account of the "white" influence was to be complete, I would have to discuss the Russians also. They were definitely Euros also and who also expanded greatly into other's territory. That was probably one of the reasons the Euros feared them. After 1492 the Euros acted as if the rest of the world was their oyster until they realized their eastern counterparts were acting the same way. I'm sure the Euros would have had little concern about the new creed if there wasn't so much industrial-military power to go with it."

"So my beginning with a glass of tea, Russian style, was probably not accidental. You are right about the western Euros; they did drink their tea from cups, and most of them put milk in it. Why the Russkis drank tea from glasses I do not know. But that's the way it was. They also made it very sweet, which I've done, even though I don't really like that much sugar."

"So now we'll talk about the people you call Russkis? Incidentally why that name?"

"Oh you should know me by now, Mary. I'm really not very reverent, and I do like slang. So I feel better calling them by a slang modification of their name based on their own language pattern. We were called Americanskis in the same mode. I could have called them Eastern Europeans, but that's so formal, and then I suppose I should have called the others Western Europeans, their formal title."

"You sound rather apologetic about your use of the term, however. Was that disapproved in your culture?"

"Yes, sort of. You see the Western Euros, after four centuries of exploiting the "others," got very goody-goody about what they called discrimination. There was a lot of self-righteous pontification about what was called "racism" and it came to be looked down upon as a derogatory term about people who had been treated very heavy-handedly in the great days of exploitation. Of course, no Euros really intended to rectify the basic exploitations. Land that had been taken away was never given back except in minuscule portions or in tricky ways, and there was never any serious effort to restore old ways of life. But one wasn't supposed to use 'bad' names. Many such were simply slang or older terms or abbreviations, the kinds of names people actually used in conversation. Thus, one wasn't supposed to call a descendant of an African a "nigger." Instead one was supposed to say "black" even though most were not because of interbreeding. Also 'nigger' and 'negro,' a later term, were simply other ways of saying "black," although with different connotations due to usage. And one wasn't supposed to say "Chink" for Chinese or "Jap" for Japanese. And for Europeans, one wasn't supposed to call a person of Italian ancestry a "Wop" which meant "without official papers" (immigration documents, that is) or a Polish person a Polack, and finally a Russian a Russki. One was supposed to be formally correct about such terms even if one wished they or their leaders would disappear from the face of the earth."

"I presume then, Pete, that you didn't follow the standard procedure."

"No, 'afraid not. I just never felt like being so deadly serious about either them or us, and not being a politician I didn't have to worry about such things. Politicians could be thrown out of office for using such terms."

"However, if you have noticed from our document so far, Mary, I was probably less ethnocentric than most of my fellow citizens, certainly the politicians. And as far as the Russkis were concerned, I thought of them simply as fellow Europeans, even if they did have a different political creed and were competing for world dominance with the other Euros."

"So, when did they go their separate way. I presume, as *sapiens* also, they shared most of the five million years of prehistory with the rest of humanity?"

Until the late Middle Ages they followed a history similar to that of the Euros. And at that time, around 1491, all of Europe was a giant royal enclave, backed by the Christian church. The Euros had Roman Catholicism while the Russkis had the Eastern version out of Constantinople. Every country had its king and queen and royal family, plus a privileged class called the nobility, supported by a great mass of peasant farmers. Then in Western Europe things began to change, including exploration of the world by sea, industrialization, and international commercialization. And by the 18th century royal heads were falling. The Euros ended up with governments that could best be called commercialized democracies, basically in the hands of political hucksters, though claiming to be run by popular will. But this claim was typical of practically all governments, including those that were run by military dictators and those which had elections with only one slate of candidates.

In the meantime, the royalty in Russia hung on and the peasants were made to carry ever-increasing onerous burdens. And though their Viking ancestors had come into European Russia by boat, they were coast huggers. Once established as the ruling class, the Vikings settled down to reap the benefits of their conquest just like their brothers in England and France, the Normans. Further, if some descendants of the Vikings still wanted adventure, there was no need to look to foreign shores. There was a vast hinterland east of the Volga which was by no means secure. The Mongols, Turks, Kazakhs, and other Central Asian peoples were still following their herds on the semi-arid plains, periodically organizing to send armies to the east, south and west. The horse nomads also would have gone north if it hadn't been so cold there. Anyway, Russia and the other Slavic nations were on the western edge of the Central Asian foment. So when Columbus and the Euros took off from European shores to the west, the Russkis went east on land. They had their hands full because there were some aggressive people out there. However, they managed, and except for the intrusion of the Hungarians, Finns, and Turks, they kept the Central Inner Asians at bay.

Some of the Euro science and technology of the 18th and 19th centuries managed to filter through to the Russkis. After all, they were then considered as kind of backward brothers to the Western Euros. In particular, they took full advantage of the evolution of the firearm so that when they started across the steppes, they had the best armed horsemen east of the Volga.

From then on the Russkies went through a historical process much like that of the Euro-Americans, and for most of the same reasons. Once the Anglo-Americans had become well established on the east coast of North America, they looked around to see what other territory was available. To the north were the Anglo-Canadians who had already displaced their Indians out of the good territory and who might be takeable. However, the Americans must have figured the price would be a little high and, in addition, most of Canada was as cold as Siberia. And then, also, they were close relatives, though generally this never was much of a deterrent in the territory-grabbing game. So the American expansionists looked south and west. It wasn't too difficult to kick the Spanish out of Florida and the French were bought out. In transferring territory, the Euros paid no attention to the claims of natives. After this transfer, it was obvious to the Euro-Americans that west was the way to go. The only important people problems remaining were the Indians of the plains and the Mexicans in the southwest. And the answer was the mounted horseman, wielding the latest firearms. And that was how the west was won.

And the Russkis? They couldn't go west because their brothers, the Western Euros, were there. But they could go north and east, and did, to Siberia and Central Asia. However, in those days it didn't contain much except vast grasslands, forests, wild animals, and reindeer herders. The south was largely blocked by Muslim nations from the Ottoman Empire to Mughal India. And to the southeast was China.

And the way to do it also was with mounted horsemen, the Cossacks, bearing the latest firearms. These horse soldiers served the same function for Russia as the U.S. Cavalry did for the U.S. They subdued the nomadic peoples, who were more or less the equivalent of the Plains Indians in North America.

The Russkis kept going east until they reached the shores of the Pacific. This expansion, like that of the Euro-Americans, was almost all on land. But expansionism ordinarily creates a hunger for more land. What Roman general was satisfied when his armies had conquered Spain and Gaul, and what Inca emperor could rest content after his armies had conquered the highlands of Peru? I remember an Israeli discussing the 1967 War in which he said the military had been amazed at how quickly the Egyptian and Jordanian forces had collapsed in only a few days. So since things had gone so very well, and since the army was all geared up, why not take the Golan Heights? Which they did in a couple of days, and the West Bank at about the same time. And this is how it must have been with the Russkis when they reached the Pacific.

Although it was frequently taught in standard histories that the expansion of the United States stopped at the Pacific Ocean, this was more fiction than fact. By the time the West Coast was safely in American hands in the middle of the 19th century, the Yankee clippers were ranging far and wide to the islands of the Pacific which could be taken without great cost in material or lives. The other Euros were also on the big water at that time, trying to get what they could before the colonial candy store was closed. So the islands of the central and south Pacific, along with their people, were parcelled out. The prize of the Euro-Americans turned out to be the Hawaiian Islands, about which they waxed very indignant when the Japanese bombed them to start World War II in the Pacific. It did not take long to feel proprietary about a newly grabbed piece of real estate. And the Americans were not even finished in the Pacific. The last major colonial takeover was the Philippines.

The Russkis, in the meantime, took some pages from the American book of conquest with their newly developed oceanic fleet. But since they had been going east on land, they continued in the same direction into the Pacific. They grabbed the Aleutian Islands and the people on them, then Alaska, and then headed down the West Coast of North America, pacifying and acculturating the natives along the way, and carrying off great quantities of furs. It has been claimed that the Russkis then had dreams of colonizing large parts of North America. Naturally they came into contact with the Western Euros, the Spanish, Canadians, and Americans, more or less in that order. The new Euros in North America were undoubtedly nervous about their big neighbor from the west and in the end the Americans bought them out. It wasn't long before the Americans regarded the Pacific as their pond. And after selling Alaska for a pittance, the Russkis turned their attention back to their newly acquired land mass, Siberia and Central Asia. A land-grabber could not sit on his laurels if he was to maintain control.

The Russkis differed somewhat from their Euro-American brothers in their methods of acculturating natives. In the U.S. the Indians were generally displaced and put on tiny fractions of their old territory, the reservations, while the rest was taken over by the Euros for farming, cattle raising, and city building. The Central Asians, being more numerous and having sizeable numbers of herding animals, were simply brought under Russian law and control but not squeezed onto tiny enclaves. There was no role for Kit Carson and other Indian pacifiers on the steppes. Further, once the Communist revolution was a fact, the Central Asian nomads were permitted to have "autonomous republics." Thus, they lived with the fiction that they were separate, independent political units (The Union of Soviet Socialistic Republics). The Russkis had no more intention of permitting ultimate control to be in the hands of the Central Asians than did the Euro-Americans with the Indians.

In other respects the Russkis went the same way as their counterparts in North America, though with minor differences. It was almost inevitable that a culture powerful enough to take over another would try to establish its own way in occupied lands. Otherwise, why go to all the effort? Also, it was dangerous to permit other cultural systems to operate within one's borders. India and Canada have experienced well the problems of multi-enthnicity, while Japan has rested easy after making over almost all their subjects into bona fide Japanese. So massive acculturation was a normal process.

In North America the Indians were forced to accept the Anglo way in almost all matters. They had to bow to the new government and particularly its system of law and order. Fundamentally, if an Indian custom went counter to an American law, the American law prevailed. Indians were hanged for ignoring U.S. private property laws, particularly regarding land. Indian warfare was, of course, forbidden by Euro-law and the Euro-style of warfare was introduced to them as legitimate. Indian conscripts and volunteers were quite welcome to fight against other Indians, Euros or Asians. And many social customs of the Indians were outlawed, usually on the grounds that they were immoral or barbaric. One which quickly comes to mind was having multiple wives, a custom a Euro-minority, the Mormons, also had to give up.

And, of course, there were hundreds of customs of the Euros which were imposed on Indians by special emissaries protected by the Euro-governments. The two most important were probably the Christian missionary and the school teacher. Most Indians were converted to Christianity by specialists operating under the Euro-umbrellas of legal power and economic influence; while the school teacher was a formal representative of the Euro-government.

One of the most significant changes was in language. The new one in America was, of course, English, superimposed at first as the language of social control and administration. Recognizing that it was difficult to get newly conquered adults to drop their own languages and learn a new one, the Indian Service concentrated their early efforts on teaching the children English. The methods were sometimes rough. Indians were forcibly sent to boarding schools where they had to speak English as well as practice other Euro-customs such as how to dress properly, use a knife and fork, and go to an indoor toilet. The main government policy was to make the Indians over as Euros. The end result was that a large proportion of Indian languages disappeared except in the journals of linguists; and most living Indians under the age of 50 became more comfortable with English than with their pre-Euro language.

And how did the nomads fare under the Russkis? They had to adjust to one major change, from a hereditary monarchy to Marxist Communism. The major conquest had been made during the monarchial period, but this came to an abrupt end during the Russian Revolution, after which all, including the Central Asian tribesmen, became "comrades" under Soviet control. There were many changes in social policy as a result.

In general, however, Russian law was imposed less forcefully than was American law on the Indians. Some social customs, like plural marriage, were permitted to continue. This was particularly important to Muslim nomads since this custom was inscribed in the sacred book. The old style of warfare was, of course, outlawed and Russki warfare approved. So just as Navahos and Winnebagos could be found as American soldiers in W.W.II, so too could Kazakhs or Mongols be found as Soviet soldiers.

The indirect Russki influence was also significant. Though there were Eastern Orthodox missionaries out among the nomads and other tribes in the early days, this all came to a halt with the Communist revolution. Comrade Marx considered religion to be the opiate of the people. So a negative view of supernaturalism was promulgated, along with other Communist dogma. Thus, the

Central Asians not only lost their missionaries, but received the creed of atheism advocated by the Soviet revolutionaries. The Russkis were, of course, acting just like the Americans did with the Indians, though with an opposite creed. That is, in both instances the dominant group imposed its own belief on those who had been conquered, Christianity for Americans, atheism for Russians.

The Communist bureaucrats, like the agents of the U.S. Indian Service, quickly recognized that older people were not easy to change from their basic beliefs. So the elderly people of both the Orthodox and Islamic religions were permitted, though not encouraged, to continue their practices. The efforts to change these groups into Marxist nonbelievers were basically on the young, undoubtedly a sound practice.

It is fitting that another of the most significant parts of the Marxist creed was introduced to the conquered people, often through force. As was generally known, Marxism condemned class differences and ownership of private property. This was undoubtedly the source of greatest conflict after the revolutionary takeover, since in pre-Communist Russia there had been enormous differences in wealth and status, with private property as the primary basis. So there was much bloodshed before capitalism, urban and rural, and private ownership of property were eliminated. The Russian peasantry probably suffered more in the purges than any other group but, of course, peasants always did.

Anyway, when the revolution was carried to the hinterlands, so was the onus on private ownership. But since Central and Northern Asia were not predominantly agricultural areas, private landholdings could hardly constitute the wealth of an individual. Instead, as with pastoral nomads everywhere, it was the herds of cattle, sheep, goats, and horses. Among peasants the land was collectivized to fulfill Marxist theory. So among herders the animals had to be collectivized.

After the revolution the Russkis went full tilt in the use of the scientific method and industry which they had taken over from their Euro-brothers. Thus, in Central and Northern Asia there was a rapid growth of manufacturing cities into which the descendants of the horse nomads were drawn, just as the American Indians had been drawn into U.S. cities. The new generation became more and more Russianized because the 20th century city in the Soviet Union grew under Russki influence. Work, housing, transport, medical and other services, clothing styles, and other aspects of life basically came from European Russia. The traditional mores of the nomads were progressively weakened.

One institution of change stood out, among both the Russkis and Euros, the compulsory education system. Once recognized that the young were most malleable, it was obvious that early brainwashing (enculturation) was crucial. Thus, school systems controlled by the state were organized and laws passed requiring children to go to classes up to a given age. The earliest brainwashing was done in the home, and by the time the child was four or five, he or she could be sent to a public school where not only the three Rs were taught, but the dogma and values emanating from Moscow or Washington. For the Euros it was capitalist, individualist democracy; for the Russkis collectivist, group-oriented socialism.

Literacy was particularly important. Children who could read could be fed much more propaganda by the state. And with the ready availability of print and other mass media, particularly television, it was obvious that teaching a single language was desirable. This was, of course, true of all great nation states. The highly cooperative, productive populations were monolingual (Japan), while the divisive, problem nations had many languages (India).

Thus, the public schools of the Euro-Americans taught English while the Russkis taught Russian. Whenever these or any other successful state gobbled up other ethnic groups, they invariably taught their central language to the conquered. And that was how the east was won.

But the Russkis, like the Euros, and particularly the Americans, did not stop after taking over the territory of tribals. Both were convinced of the righteousness of their systems and did all they could to spread them, even in areas they did not directly conquer. The major technique was to encourage men from the upper class of the colored countries to study at their universities where their values and version of history were taught. Thus, the countries of the colored man's world came into the industrial age usually with one version or the other, collective state control or individualistic capitalistic control. The rivalry between the systems led to many revolutions. Japan, the Philippine's, Thailand, and Malaysia went capitalistic while China, North Korea, and Indo-China went collectivistic. Most of the post-colonial countries of Africa and Latin America also split along these lines. A late bone of contention to the U.S. was Cuba, 90 miles from the bastion of capitalism, which went Marxist. Few countries went on independent paths. Both the Russkis and Americans tried to induce such countries to go their way by doling out aid according to compliance with capitalist or collectivist principles.

But unlike their Western brothers, the Russkis failed to keep their great empire together. Although the Soviets accomplished much in industry, technology and warfare, they never managed to get their Marxist system going well economically. So in the last decade of the 20th century their economic and political systems collapsed. And a large number of the peoples they had gobbled up in their expansionist days demanded and got back their independence. Most of the central Asians got their own nationhood back. And that was how the Southeast was lost.

4

The Eurogene

I was settled comfortably in one of the easy chairs, smoking my pipe, thinking about why I had taken up the habit again. Both video screens were dark which at that moment pleased me. I was content with my own thoughts, not yet ready to share them again with the electronic device of an extraterrestrial. I watched the smoke billow up and spread, remembering how pleasant it had been to study the ever-changing forms of tobacco smoke. I even tried a couple of smoke rings and was pleased with my success. The taste was also pleasant.

I remembered clearly my early mixed reaction to the habit. I had been addicted to cigarettes for fifteen years or so, and though I enjoyed them, particularly the first thing in the morning and after meals, I would frequently get annoyed with myself for smoking too much. At such times the taste would be distinctly unpleasant, and frequently I would stub out the just-lit cigarette in exasperation. Of course, after a while I would light another.

I knew also that cigarette smoke was not pleasing to others, particularly to my wife. So when she suggested I try a pipe, I was ready. Then began another fifteen or so years of my love affair with tobacco, though with a far greater payoff than with cigarettes. My wife liked me to smoke the pipe, both for its aroma, and because she thought it made me look professorial. Also, the aroma frequently drew compliments from others. And finally, I enjoyed the taste of aromatic to-bacco smoke much more than that of cigarettes.

But then the great tobacco cancer scare came. The pipe rarely caused me to cough, though I did sometimes get a sore lip, from which, it was claimed, cancer could result. I was in a health improvement mode in those years, my forties, so among other changes, I stopped smoking altogether. I still remember the day I took all my pipes into the garage, lay them in a row on the concrete floor, and shattered them one by one with a hammer. I still feel it as a kind of betrayal, even to those inanimate objects. But I did drop the habit.

In the interview room I fingered the gnarly briar bowl. It was a good feeling. And my thoughts jumped forward to the present, to when I had resumed the habit. I was 96 years old, and in much better health than I had expected to be. I had of course received good medical care in my later years, having most of my teeth replaced first by implants and then dentures, as well as getting hip replacements for arthritis. Moreover, my veggie diet had not hurt.

So one day after Takeover, while I was feeling a little depressed at all the changes that had taken place, I was studying a little pile of odds and ends that an earthling was peddling. Among them were a couple of briar pipes which I had bought on impulse. I found tobacco and matches with little trouble and before one could say *"Homo sapiens sapiens,"* I was smoking again. I suspect I was doing it partially because it was a link to that old world when man was dominant, which seemed like an eon.

The ambiance of the room began to change, first with music, then a low humming, then subtle clicking, and finally a flickering of video lights. I kept my pipe lit while watching the screens. The surrounding lights turned into green and Mary came to life with the words, "Hello Pete, how are you feeling?"

"Oh fine, no complaints." I watched my words roll across the I-screen. The "I" I learned later, meant "informant."

"I see you have taken up a new habit, the earthling custom of smoking. I don't think you were doing that before."

"No I wasn't for many years before you Atierrans took over. Though I did smoke in my younger and middle years."

"Unusual custom, I must say. I cannot recall having heard of it on other worlds. Of course, many other substances are used elsewhere to slow down or speed up the body functions, or to help individuals through crises such as death transition, or simply to promote a feeling of euphoria. But not ingesting the gases of burning substances."

I knocked the ashes from my briar into a convenient dish. Then I began the ritual of refilling the bowl and lighting up. "It is a custom we picked up from the American Indians. In fact, in the later years of the 20th century we referred to it as the "Red Man's Revenge" because it was discovered to produce cancer. Euroman adopted quite a few plants they had domesticated, and one was *Nicotiana tabacum*. The weed was smoked, chewed, and drunk; and incidentally, the forms of use that spread around the world originated almost entirely with the Indians. Euromen found the Aztecs smoking cigarettes, the Indians of Cuba smoking cigars, and the Indians of what was to become the United States smoking pipes and chewing the tobacco. An incident we used to tell our school children, probably a myth, is that water was thrown on Sir Walter Raleigh the first time he was noticed smoking a pipe. He was one of the pirates the English had unleashed on the Spanish to get whatever loot he could get his hands on. Anyway, like most other conquerors, he took what was interesting from the conquered, thus helping pass on the "Red Man's Revenge.""

Mary responded, "Interesting story. Your Euromen were certainly busy fellows. But I'm surprised to find you taking up the habit again. Aren't you concerned about getting the illness?"

Before answering, I deliberately took a long draw and exhaled. "No, not really. You know I'm quite old by earthling standards, 96 years, and I seem to be in surprisingly good health. And then it is still generally agreed that pipe smoking is one of the least harmful ways of using the plant."

I paused a moment, concerned that I might anger my interviewer. But since there had been no serious problems of rapport so far, I decided to say what was on my mind. "Probably another reason is that these days I do sometimes feel a little depressed about the loss of so many of the old ways. It's something our brother social scientists, those we called sociologists, used to call "anomie" or a feeling of emptiness because of a loss of so many predictable cues." Then I quickly added, "It's not that the Atierran way is bad, just that it's so different. And also you know we earthlings now have little control over our lives. Any little old custom gives comfort."

Mary was quiet for a moment, but then responded in her normal succinct way. "I see. You are saying you need some of the old ways to feel at ease? Well, I understand. As you know, we Atierrans try to create as many conditions as we can to make you feel at home."

"I do, and I appreciate it."

We both stopped momentarily, then Mary resumed. "So I guess if you don't mind, we could go on from where we left off yesterday. Is that okay?"

"Sure, let's do it."

"So Euroman spread throughout the world in 200 to 300 years, right? Sometimes he swallowed other cultures, particularly those of less complex societies, but he brought great changes everywhere, even to old civilizations."

"Yes, whatever else one could say, one would have to admit that the entire world was changed profoundly by the light-skinned adventurers from the peninsula of the main Old World land mass which we called Europe.

"Euromen were not the first expansionists the world suffered through, of course. When they had the power, men usually took over the territory of others and whatever else seemed worth taking. And though various barbarian types were sometimes successful, including the earlier Central Asians and the Germanic tribes, the most successful expansionists were the urban societies. In historic times we can count many, such as those of ancient Mesopotamia, Egypt, China, Persia, Mexico, Peru, and Rome. All, however, were land based and consequently limited. Some were so localized that they did not know of the others' existence. But only the last great expansionistic wave was worldwide, the one which came out of Europe on the ships of the palefaces at the end of the 15th century and their eastern brothers by land.

"In any event, it is inconceivable that one could understand how man got to be the way he was when you Atierrans took over without knowing about Euroman's effect.

"The hairless, tool-using biped we know as *Homo sapiens* had indeed evolved fully before the expansion of Euroman, but most of the particulars of his way of life occurred after 1492."

"All right, agreed. And where to begin?"

"Why don't we begin with the physical changes because many took place when the European explorers got to the new territories. Moreover, many of the world's attitudes were shaped by the physical differences between Euroman and the others, the ones we will call 'colored.' Euroman was called 'Whiteman'."

Among many new sciences that emerged in Euroman's culture, what we used to call Western European Civilization, was one that was concerned with how physical changes took place. The science of genetics really got started in the 20th century and gave mankind more of an understanding about biological changes than anything had before. In particular, it produced theories about how new physical traits occurred and how they were spread from individual to individual and group to group.

Most earlier peoples had known that sex and reproduction were related, that males and females had to get together to have babies. And that the kinds of babies could be controlled by seeing to it that only certain kinds of males got to certain kinds of females.

One of the important discoveries of mankind, long before Euroman's day, was the domestication of plants and animals. This entailed controlling their life cycle for human benefit. One of the techniques that paid off was the elimination of undesirable animals and plants by killing and eating them or castrating the undesirable males and then eating them or using them as draft animals. As your anatomists must know, the males of bisexual animals have special breeding organs, called balls or testicles. Being in neat little sacs on the outside of the body, they could be removed easily. And that's what the animal breeders of the world did to keep undesirable animals from breeding. Men even got around to doing it to some of their own kind, to serve as harem guards or to become opera singers.

But that's all men seemed to have learned about reproduction and the inheritance of physical traits until the science of genetics emerged in the late 19th and early 20th centuries. Then a series of discoveries revolutionized thinking. One was that in the fluids produced by the testicles and the female equivalent, the egg, there were invisible units called genes which controlled the physical appearance of the newborn. Moreover, these genes sometimes changed, producing new variations. For instance, the normal eye color of *H. sapiens* is brown, and yet a variant, blue eyes, appeared in northern Europe and spread widely throughout the world. This is thought to have been a result of a mutation or variation in the gene. But the variation would be of little importance if it had not been spread so wide. And how did this take place?

First, it was a consequence of basic mammal behavior that sexual privilege goes to the dominant. Males fight to control territory and females. Monkeys and apes fit the pattern well, so it is not surprising that the super ape, *H. sapiens*, continued the same behavior. But being an accumulator and a communicator, he generally got his mate/s according to the amount of resources he controlled and his ability to communicate. This represented human power as contrasted to horn power for sheep and fang power for wolves.

Anyway, man got his mate in his own society by parlaying his wealth and position more than his physical prowess. But he sometimes even used his muscles, usually for show only, but for fighting if necessary.

So it should come as little surprise that the conquerors got access to the women of the conquered. Wherever Euroman established himself in a colored group, he took some of their women. Since the initial explorers were almost all males, the unions were white male-colored female. And even when female whites were brought to the newly conquered lands, they were almost always off limits to the colored males. This pattern continued into the 20th century. Though generations of white males had access to black females, a sexual liaison between a black male and while female could still easily lead to violence against the black male. Euroamericans, who hung on to slavery the longest were notorious for their lynching of such black males.

Euromen did not usually have to use violence to get colored females. The native males not infrequently provided them with mistresses, usually in an effort to buy them off. This rarely worked of course. The Euromen took their pleasure with the women and then proceeded to take over the territory, natural resources, trade items or slaves anyway. And even when Euromen were not offered women as bribes (not brides), they did not necessarily have to resort to force. Euroman ordinarily encountered colored people as a conqueror. Although at first the landing party might be resisted and some people even killed, each successive landing party brought more, better armed Euromen until they were established.

So in the beginning the newcomers were sometimes presented with sexual partners. Who knows how often these women were pleased with this arrangement? A documentary film was made of the first expedition of Euroman into central New Guinea, at which time Melanesian women were offered to the Australian explorers. Some fifty years later the same women were interviewed and asked their feelings at the initial encounter. They had been awed by the white men to whom they were presented, fearing that they might be eaten. These Melanesians had practiced cannibalism themselves, so it was not so strange that they might consider that possibility from powerful strangers. The women were happy when they realized that the Euromen did not want to do anything worse to them than their own men did. A number of them begat children from the unions, contributing persons to the new type of *sapiens*, the half-breed.

So in general, white men had little difficulty convincing some of the colored women to be their mistresses. The men were taking the dominant male role and fulfilling their Darwinian destiny by spreading their seed.

The first Euromen treated the first tribal people they encountered either as impediments or sources of exploitation, and this despite the fact that they were basically the same physical type. We now know that all human beings belong to a single species. This means they can interbreed and that there is a physical attraction between the sexes, no matter what other differences there may be. Though we also now know that there is a very close relationship between our species and the chimpanzee, we never did find a clear-cut case of men and chimps mating with each other.

Euromen were hairless bipeds with language and tools, as were the natives whose territory they were invading. So one might ask, didn't this make them more considerate than they would have been with a new species of animal? Certainly it did not stop the invaders from taking over the territory and/or irrevocably changing the way of life of the natives. Spaniards began by wiping out all the Indians of the West Indies just as other Euromen wiped out the dodo, Carolina paroqueet and passenger pigeon.

However, there was a difference with natives because of the perceived kinship. The invaders, seeing a physical affinity, searched for sexual partners. So while displacing or subduing the natives in general, they fucked their women. And they left behind a new type throughout the world, called variously half-breed, mestizo, metis, Eurasian, Anglo-Indian, mulatto, and Reheboth Bastard. And thus the blue eyes and "red" hair of the north Europeans was spread throughout the world. No wonder the Chinese called them the red-haired devils, and more recently Ali Mohammed called them the blue-eyed devils. To the natives they were certainly devils, and among them there were a fair number with light-colored hair and eyes.

People interbreed no matter what the social distance. I was impressed by a very different physical type I encountered once while on a field trip in Trinidad. They were very handsome with medium tan skin, black wavy hair, and well-balanced features. I learned that they were of Chinese-African ancestry. In its turbulent history after the whites came, African slaves had been brought to the island and later, after slavery had been abolished, indentured laborers from China and India were added to the melting pot by the British. The original inhabitants, the American Indians, had been eliminated by that time. In the early days the indentured laborers were almost all men because they were the most useful to the plantation owners. However, the men wanted women and took what was available, Afro-Caribbeans.

The women encountered by Euromen often looked different, but were obviously human. Once that was established, the relationships started as a simple matter of satisfying the sex drive. But because of their humanity, it did not end there. Man has from the beginning been a cultural animal, and an important part of his culture has been the social rules. So while Euroman had access to the local brown-eyed maidens, the daughters and wives of the conquered, the brown-eyed men rarely had access to the daughters of the blue-eyed devils. Thus, the mixed breeds of the world were almost all products of Euro fathers and native mothers. The population genetics researchers came up with a term for this process, gene flow. It is when genes go from one group to another when groups come into contact. When that happened, the physical characteristics, blue eyes and blond hair, followed.

The Eurogene also spread as a direct result of conquest and cultural destruction, especially of the tribals. When Euroman conquered a given area, the native culture invariably started to disintegrate. The Americas, Australia, and the Pacific Islands are classic cases. Epidemic diseases quickly took their toll. Then the process of displacement went into full swing, Euroman constantly pushing the native groups toward the fringes, killing many in the process. Eventually the surviving natives were forced to accept whatever small parcels of territory Euroman did not want. In the meantime the way of life of the natives was replaced by the Euro-way. Among the other disasters, the new Euro-foods were worse for the health. The population of the natives began to decline.

At the same time, Euroman and woman moved into the new territories and began to beget at a great rate. The earliest displacers were rural people who generally have large numbers of children. Such was the case with the Euro-settlers. The increase in population of Euroman started to slow down only when the natives had been eliminated as competitors and white men began to settle in the cities. But by that time the tribals had been reduced to insignificance by the Euro-breeders.

The direct consequence of this gene flow has been the spread of the Eurogene while the indirect consequences have been social.

Did Euroman really differ from others in physical appearance? There were indeed some characteristics which still affected our lives when Takeover occurred. Undoubtedly the most significant was Euroman's skin color, which turned out to have the most far reaching social consequences. Although there was some range in color in the groups of Euroman, compared to most other people of the world, he had a lighter skin. Physical anthropologists generally believe this was due to Euroman's home habitat. In general, northern Europe is cloudy; and the sun is weak as compared to the earth's middle belt, the tropics. There was no biological advantage to darker skin color in Europe, so it is assumed that when mutations occurred, they continued through reproduction, leaving Euroman with a pale skin.

It seems logical to assume that the original skin color of mankind was dark, since all his closest relatives, the monkeys and apes, have dark skins. However, they all live in the tropics or subtropics.

Two other color changes seem to have occurred in the population of Euroman, again from presumed mutations. Blue and green eyes seem to have evolved from the standard brown eyes of most of the world's peoples. And again, dark-colored eyes are the usual for monkeys and apes. *H. Sapiens europus* is the only significant exception in the primate order. Also, light colored eyes were just as functional in the less sunny climate of north Europe. The other variation was in hair color, blondism, and again this was okay in Europe.

There also were some minor differences in genetic makeup. The most typical nose form of Euroman was narrow, while the people of the rest of the world generally had wider noses. And the lips of Euroman were thin in contrast to those of most others. In addition, Euroman had more body hair. In this sense he was more like his ancestors, the hairy apes. No one knows exactly why man became relatively hairless though, of course, he was the only primate that wore clothing and used houses to protect himself from weather extremes. Hair was less important for temperature control. We anthropologists went out among other peoples for our studies much more than did those from other professions. In fact, we always prided ourselves on our fieldwork and looked down on what we called "armchair theorists." Thus, non-Euromen frequently got a better look at us than they did of others of our subspecies. And they watched us closely.

I remember going about by foot at first in Trinidad. Later I got a car. My wife who is also of European descent, was often with me on these walks. One day I was talking to a very black Trinidadian in the open marketplace. He was of African descent as was about half the population of the island. Our conversation got around to transportation and he said, "It be good you have auto now to take your woman about. I used see you two walk about and think if I have woman as white as that I not make her walk but take her about in style."

To reiterate, the natives of wherever, who were usually what we called colored, studied us closely. I remember such an experience in a small hotel in upcountry Thailand. After my family and I had settled down in our hot room and were taking off some outer clothing, we saw the manager watching us closely through one of the barred windows. "What do you think he wants?" I asked my wife.

"I don't know, but I certainly don't feel comfortable. Ask him."

I did and he said he didn't want anything, and remained where he was.

In our training we anthropologists had been taught not to alienate the natives, so my wife and I tried to ignore him. But she could only take it so long, and finally said, "You have to do something about him. I just can't relax with him staring."

I then said to the man, "Why are you standing there?"

"Just watching. We don't get white people here often."

On another field trip I had enrolled my eight-year-old son at the local Lao school where he learned the hard way about differences in appearance. He was having trouble being accepted by the local boys who were almost all East Asians. He came home crying several times because the boys, he said, were being mean to him. He said they threw rocks and pulled his hair. I understood the rock throwing, an unfortunate characteristic of boys in many cultures though I suppose natural for a species which began the long road to dominance by throwing rocks at other animals. I didn't quite understand the hair pulling, however since my son did not have long hair.

He showed me where the Lao boys had pinched and twisted the hair on his forearm.

"Why would they do that?" I asked.

"Aw, they don't think it's real. They want to see if it will come off."

I puzzled about this until the proverbial light came on. Apart from a few remnant groups such as the Australian aborigines and the Ainu, Europeans were the hairiest of the species while East Asians had the least body hair. In old paintings one sees the Chinese sage fondly stroking the half dozen hairs which constituted his beard. Many American Indians, who were believed to be direct descendants of East Asians, could get rid of their extra facial hair by plucking out the few they had. A German or Italian who tried that would be in for a painful ordeal. It was no accident that one of the best known oaths of the followers of Mohammed, who were genetically white men, was "by the beard of the Prophet." Such an oath would have been ridiculous if the founder had been Japanese or Chinese.

Anyway, there was my son having the hair on his forearm plucked because he was a hairy white person. Actually he had quite a lot of body hair even for a European.

In any event, Euroman went forth on the seas of the world with a slightly different genetic makeup. In general, geneticists believe that all members of the human species share more than 90 percent of their genes, but the less than 10 percent of difference was enough to have created many difficulties in the post-Euro world.

There seems to be little doubt that the most important difference was skin color. European explorers were almost always lighter than the people they met. Euroman was not really white, but rather a shade of tan. The only really white people are albinos who can come from any subspecies. They totally lack pigment, thus having white skin and pink eyes. However, all normal Euromen had pigment, albeit less than the other peoples of the world.

No matter, this distinction was not made in the popular classification systems of everyday languages. Euroman became the "whiteman" and the other peoples of the world became "colored." All kinds of unusual criteria were used to make these distinctions, most of which still continued to bug mankind at the time of Takeover. In our minds we ended up getting negroes or blacks, a skin coloration which made more sense than most, even though many natives of sub-Saharan Africa were a dark brown. Then we got Red men or Red Indians. They got their name from facial paint they frequently put on to meet the first Euromen. Their skin color was a shade of tan or brown varying by tribal group. Then we got yellow men, how I know not. These are the people of East Asia, the Chinese, Japanese, Koreans, Vietnamese and some others whose skin color is a shade of tan. Logically, red men and yellow men should have had basically the same skin color since the Indians are thought to be descendants of the East Asians.

In any event, popular concepts of skin color were ridiculous from the beginning and remained so until Takeover. One of the peculiar occurrences was that mankind was so deeply brainwashed about skin color that ideas about them were often believed no matter what could be clearly seen. Probably most college students in my teaching days actually believed Euroman was white, while men from China and Japan were yellow, and Indians were red. I used to hold up a blank piece of writing paper next to my face and ask, "Is this the color of my skin?"

My students would grin or titter and we would go on to the next topic.

While skin color outside Europe varied quite a bit, it was generally black or some shade of brown. Thus the term, colored, came to be widely used as the opposite of white.

Also, skin color came to be the main basis for an attitude that lasted throughout the era of Euroman, what came to be called racism. Since the conquerors everywhere were light skinned while the conquered were of color, white came to be thought of as good while dark was bad. This idea became so pervasive that even when extensive interbreeding took place, particularly between whites (men of European ancestry) and blacks (women of African ancestry), the lighter-colored, mixed breeds were favored in marriage as well as in the job market. And mixing became so pervasive in Euro-America that the great majority of what continued to be called blacks were no darker than a European from the Mediterranean region. We came to have politicians and other professionals at the end of the 20th century whose sole claim to being black were slightly thick lips and wavy, black hair. By then there were some advantages, particularly for politicians, to be able to claim African ancestry, even when the person's genes were 90 percent white.

A preference for light skin color was not solely a characteristic of Euroman, though it is a little hard to know how much his preference affected others. In any event, in the civilizations of Asia and the Middle East, lighter skin colors were favored, particularly in the marriage mart.

Back in the days when I was doing much fieldwork, I did a study of marital advertisements in Indian newspapers. Since we *sapiens* were particularly concerned with the physical appearance of our females, this attribute often was listed in the ads for women. Men wanted a beautiful girl with light skin. Other qualities were listed less often. In what we called the Far East, lighter color also was desired in women. Both the Japanese and Chinese upper-class women, and particularly their professional hostesses and entertainers, powdered their skins to be milky white. And women who had to work in the field wore special widebrimmed hats with veils around the rim to keep out the sun's rays. I have seen such workers with heads wrapped so completely, they looked like living mummies. The temperature did not seem to matter as much as the need for keeping the skin light.

An interesting contrast with respect to exposure to the sun existed among the affluent who could afford to use European style hotels. Swimming pools were invariably built as part of the elaborate hotel complexes in international hotels. And though the local beauties would sometimes display themselves in brief suits, and even go into the water on occasion, they did not sun themselves. This was a leisure-time activity that milky white Eurowoman had developed to perfection. In the cloudy north they tried to get whatever sun they could, so when they got to sunnier climes and became urbanized, they took up the pursuit of the sun even more seriously.

As long as men and women continued to work in the fields in North America, they kept their heads covered. Our farmers and cowboys were practically never seen without their widebrimmed hats while their womenfolk wore bonnets. They may, of course, have recognized the debilitating effect of the sun on a light skin, but this is only recent medical knowledge. But when they became factory and office workers in the cities, they discarded their head coverings and moreover took to tanning themselves as a sign of leisure. In fact, the pursuit of the suntan became commercialized in the development of tanning studios. In the meantime, the Asian ladies, even the urbanized, did all they could to keep their skin light. To them a tan was a sign of the peasant way of life.

It may seem odd that Euroman/woman based some of her ideas of superiority on lightness of skin color while still pursuing the suntan. However, it has long been known that cultural concepts are never perfectly compatible with one another. In fact, mankind is quite capable of maintaining contradictory ideas.

Euroman was certainly not the first to dominate others. But more often than not, other civilizations thought their superiority was a consequence of their superior ways rather than their physical appearance. A full-fledged Egyptian, Greek, Roman or Chinese assumed he was superior because he spoke the traditional language and followed traditional customs. Sometimes, as in India, physical appearance was a consideration, but even there the customary behavior was more important than skin color. A very dark southern Brahmin was still a Brahmin, while a lighter-skinned Chamar (leather worker) in the north was much lower in status. The emphasis on skin color as the primary criterion for social status seemed to have been Euroman's particular contribution to status differences. And even there it was primarily the north European who set the standard, since the people of the Mediterranean did not differ so much physically from the people they conquered. To the day of Takeover it was claimed that one achieved acceptable status in Latin America by learning how to speak Spanish properly and following other customs that came from Europe. One was a low-class Indian for speaking an Indian language, dressing like an Indian, and having other Indian customs. By contrast, a "nigger" in the U.S. was a person with some amount of African ancestry, much less what customs he followed.

If it is true that Euroman set up skin color as the primary criterion for racism, it is just as true that the colored of the world reacted strongly to the pallid hue of Euroman on their own. As mentioned, in all likelihood in the civilizations of Asia and the Middle East lightness of skin already had a positive value, particularly for women. In most of those countries, besides being characteristic of the dominant groups, it was an indication of higher social status of people who did not work in the fields. But to most "others," the pallid hue of Euroman was more surprising and difficult to deal with.

Not surprisingly, it was the very dark people who had particular difficulty assimilating the idea of powerful strangers with light skins. These very dark brown- or black-skinned people had difficulty imagining that such a light color could truly be a human variation, so they fell back on the beliefs they had held before Euroman arrived. A pure white skin is almost an impossibility since blood has to course through the veins. As mentioned the closest we find in living humans is in albinos, the consequence of genetic mutation. Otherwise, light skin is an indication of illness or death. Thus, logically some native peoples came up with the idea that whites were dead ancestors or ghosts.

Nobody went further with this idea than the Melanesians of the South Pacific, and since they seemed to be as materialistic as the white intruders, they also evolved the idea that these ancestor-ghosts, their living descendants, were bringing material goods to them.

This ghost-making evolved through the usual tribulations of tribal people. After Euroman had established himself on the coasts, the natives at times tried to resist the interloper directly and suffered the consequences. Their spears and bows and arrows were no match for the white man's firearms. So they moved to the second strategy devised by assaulted groups, they created a new creed and ritual. They came up with the cargo cult, a ritual based on the idea that the whites were their ghostly ancestors bringing the goods of a technological civilization. The strategy was for the natives to cease traditional work, do special dances, and try to get the pale ghosts to bring in their cargo-laden boats or airplanes. However, by that time, the white administrators were hardly pleased with their new colonial subjects and had the prophets of the new cults carried off to mental institutions or jails. So, in many places the Melanesians modified their beliefs. They decided that while the crews of ships and airplanes truly were their ghostly ancestors, the administrators and military were not. The undesirable whites would disappear and there would be a materialistic paradise.

Though the cult may sound bizarre, it was widely believed by anthropologists to be a type which occurred frequently when weaker people were subdued by stronger ones. Some claimed that both Judaism and Christianity began as such messianic revitalization movements to give the downtrodden Hebraic Jews and later Jew/Christians hope in the face of oppression from Egypt, Assyria, Babylonia, and Rome.

But back to the Melanesians. As with most such movements, their cargo cults did not work. No cargo ever came, and the white intruder-oppressors (Germans, Dutch, English, Australians) did all they could to suppress them. The cults died out and after a few decades the culture of the Melanesians was radically changed toward that of the palefaces. They got democracy, Christianity, and clothes.

The blacks of the west coast of Africa came up with many versions of similar cults, relying on ritual to get the white man out with or without his goods once they learned that direct resistance would not work.

It seems hardly an accident that Africans and Melanesians tried hard to deal with the white menace. They suffered extensively at the hands of Euroman. The system of chattel slavery of Africans began with the first Portuguese explorers and lasted for 350 years, one Eurogroup picking up the practice from another. Besides all the violence perpetuated on the Africans during the slave period, many Euro-nations ended up with populations of millions of their descendants, along with continuing systems of discrimination in many countries, particularly the United States.

It is less well known, but a similar system of forced labor, combined with kidnapping, occurred in Melanesia. It was called "black-birding," the British hauling off Melanesian villagers to work their coconut plantations.

Anyway, the exploitation of blacks and their reaction continued right up to Takeover. Although the slogan "black is beautiful" was coined in the era of black power in the U.S., in actual fact lighter-colored blacks were still preferred over darker-colored ones by them as well as by whites. This became the primary racial problem of the 20th century.

The other color variations of Euroman had lesser effects, though also continuing until Takeover. Blue eyes and their genes were carried into the brown-eyed world of the colored peoples. Despite the fact that blue eyes were inferior for filtering the sun's rays, they became socially desirable. By the time we got through the high point of Euroman's dominance, the early 20th century, blue eyes were in all world populations, spread by the sexual activity of the palefaces. And when mass media entertainment became established worldwide, the white bias of Hollywood was disseminated everywhere. An inordinate number of superstars, both male and female, had blue eyes. This was one of the reasons of course why they had been selected. Euroman even came up with a political/military movement based on the supposed superiority of blue eyes, blond hair, and light skin. That was the concept of Nordic supremacy, advocated by Hitler and the Nazi Party. This helped bring about humankind's biggest war, World War II. Ironically Hitler had neither blue eyes nor blond hair, though he was white.

Blue eyes became so popular in the United States that cosmetic techniques were developed to make brown eyes look blue, though not the reverse. If a female wanted to enhance her beauty she got blue lenses. Our most highly admired female actresses in the 20th century were blue-eyed blonds. Green eyes, a variant of blues, were valued slightly less.

Blond hair also was spread by Euroman and had a very similar effect. A truly handsome male or beautiful female had to have both. A variant of blond hair was red, which like green eyes, was valued somewhat less than blond. Cosmetic techniques evolved to give blond hair to females not so endowed, either by dyeing or by using a wig.

As mentioned, at least one great culture, China, named Euroman by his hair color, referring to him as the red-haired devil. Devil referred, of course, to the hell he raised when he got a foothold in China.

Another minor characteristic of European hair had world- wide repercussions, its waviness. The head hair of a Europerson generally had a wave in it which became accepted in many places as the ideal. Eurowoman also added to the natural waviness by the use of curlers. As a result, people whose hair was straight often went to considerable trouble to make it wavy. And what would be more natural than using the curling devices already developed by Europeople.

A reverse procedure came into existence among persons of African ancestry, particularly in the U.S. Africans not only did not have the straight hair of East Asians, they had hair that was literally frizzy. To move toward the ideal of

43

wavy hair, they had to eliminate some of the kinkiness. This was done extensively in the U.S. before the era of black power, either with chemicals or electric de-curlers. It was called "conking," presumably from the slang meaning of conk, a blow on the head. Fortunately for black females, when the era of "black is beautiful" came in, some decided that "frizzy was fine." New styles evolved by which a person of African ancestry could use the natural kinkiness in a hairdo. Afro-Americans came up with the "natural" which was super frizzy. Wavy or de-frizzed black hair came to be a distinguisher of generations. Older persons, sometimes referred to as Uncle Toms or Aunt Tomasinas, had straightened hair while younger, militant blacks flaunted their "naturals."

Although Euromen/women generally had wavy hair, variations included straight and frizzy. The transfer of genes had gone on for as long as we know, and it went on at a great rate since Euroman went on his international prowl. So practically any trait could be found in any population, including frizzy hair among Euromen/women. I knew a couple of women of European descent, one from Eastern European Jewry, the other from Scotland, who had very frizzy natural hair. The Scotch-American woman had gone through an upper-class version of conking to try to get some of the kinkiness out of her hair. It would be wavy for a day or two but go kinky with the first dampness. As a child she had been accused of "having a touch of the tarbrush," an expression used early in the American south for having some African admixture in one's ancestry. The Jewish-American woman also had gone through a straightening process for most of her life which she claimed had been a real pain. But when the Afro-American "natural" became respectable, she happily teased her frizzy mop into that style.

The only other physical trait which caused some repercussions was the bony structure of the European nose. Euroman generally had a narrow, straight nose in contrast to the wider, flatter nose of most colored people. Actually, the classic straight nose was typical of northern Europe while the people of the Mediterranean had somewhat wider noses. But even more significant was that in the Mediterranean region a hooked nose occurred frequently. It was sometimes called a Roman nose though many other people of the region, including Arabs also, frequently had it. So when the Jews, who were close genetic relatives to the Arabs, were driven into northern Europe, they carried the hooked nose along. And once they came to consider themselves as northern Europeans, they turned to cosmetic surgery to make themselves look like Euroman/woman. It was quite the rage in the U.S. in the 20th century for upwardly mobile Jewish families to have the hook or bump removed surgically, especially on their children. The girls in particular could then do better in the marriage exchange system.

5

The Tool User

I watched the robot with considerable interest. On earth we had, of course, developed some pretty fair robots by the time of Takeover, but they were still simple machines compared to those of the Atierrans. I really should not have been surprised since so much else about the aliens had been far advanced over what we had.

The robot was a model of concentration as it tinkered with a circuit of Mary's screen. I was particularly interested in the flexible antennas as they moved back and forth, evidently picking up vibrations. The manipulation of its multiple appendages also was interesting. They reminded me of the antennas of insects, and I knew they were fulfilling a similar function. The appendages were busy manipulating the parts of the circuit, but ever so gently. Once in a while, when one of the appendages lightly stroked some section of the circuit, an arm would spark brightly for a microsecond and pull back. The robot had done some kind of welding or electronic joining, I figured. The lights began to flicker, followed by a lightening of the screen. The robot then flowed backward a little way and waited. The music of Petrouchka came on. The robot collected the various tools with its multiple arms, placed them methodically in the carrying case, then moved off in its flowing fashion. I admired how neatly four of the appendages were folded while the other four carried it smoothly across the floor. So logical, I thought. If bipedalism had given *sapiens* an advantage over quadrupeds, it still had been limiting to have only two arms. Making a robot with eight appendages on the model of the octopus made sense. Mary spoke, "You find our robot interesting, I see."

"Yes, I compare it to *sapiens* and even the robots we developed. We were locked into a way of thinking that came from our bipedalism, you know. Although our robotics engineers developed robots with more than two arms, whenever possible they tried to come up with two. It was natural, I'm sure, just like the decimal system which was derived from our ten fingers; but having more appendages as a standard feature, gives your robots a real advantage."

"In what way?"

"Primarily for tool use. Having six appendages for manipulation makes a big advantage over two. I am assuming that at least two of the appendages have to be used to hold the robot up."

My musings carried me off on a tangent. "Of course multiple arms could have facilitated other activities. I remember seeing an erotic painting by a Japanese artist once in which a man was trying to seduce five women at once. He was using everything he had, but even so was contorted. Imagine what he could have done with eight arms/legs. Also, of course, multiple arms were a standard feature of Hindu deities. They symbolized the super power of gods."

I sighed. "But we have to recognize our evolutionary heritage of four-footedness evolving into two-footedness/two-handedness. And after all, it did give us an edge over other species for five million years, more or less."

"Tool use, tool use," the words flickered, Mary seemingly musing. "Didn't you anthropologists use that as a definition of humanness, those who used tools being considered to be of your kind while the non-tool using animals were thought of as sub-human?"

"Some other animals did use tools, but only in a limited or specialized way. *Sapiens* is the only one that literally went amok in creating varieties of tools. So in general tool use can be used as a criterion for humanness. The part of anthropology we called archeology used tools as its main stock in trade. Whenever the fossilized bones of an ancient *Homo* type were found, the searchers invariably looked for tools, generally of stone. Only then would they be willing to consider the fossil as a real ancestor."

"Interesting. I have heard little of stone being used for tools since we came. I presume mankind changed in his use of materials."

"Exactly. As a matter of fact, we had a series of evolutionary stages that we used to describe mankind's improving technology, developed just a little over a hundred years ago. The evolutionary stages consisted of the ages of chipped stone, then polished stone, then bronze, and finally iron. When Euroman went forth, we were still living in the age of iron. Later of course, we got into other metals and of course, plastics."

"So this stone and metal was used for your tools, the devices you used for exploiting your environment, is that right?"

"Yes, we used them primarily for that, but for other things also. It is these tools that our ancient forebears used with their newly freed, sensitive hands. Man's conquest of the earth was the consequence. Other animals had specialized body parts, but each had a limited function, while our species could make tools that would compete with the specializations of all the other animals. At first we stayed on land, fighting it out with the largest of the other creatures, but eventually we got so good at using our tools we challenged the mightiest creatures of the oceans. If we had not slowed down, we would have wiped out many more of the great creatures before Takeover, the cats, bears, wolves, rhinos, elephants and whales, to name only a few. We even took a kind of pride in the varieties we eventually included in what we called 'the endangered species list,' those we saved after having brought them to the brink of extinction."

"You people certainly sound like fierce animals. Do you think that is an honest appraisal?"

"Yes, I'm afraid so. Although many of our societies, particularly those who claimed to be civilized, advocated peace, we perpetrated more violence on our own kind and others than any other species known. We liked to claim that our folk heroes advocated peace, such as the founder of our principal Euro-religion, Christianity. Among other titles, he was called the "prince of peace," yet all kinds of wars were waged in his name. This is not to mention endless carnage against other creatures for 2,000 years. We had large factories, called slaughterhouses, for killing and dismembering other animals."

I mused about my outpouring for a moment, thinking it was a little strong. Then to soften it a little, I went on, "I'm not at all sure other creatures wouldn't have killed just as much if they'd had the chance. Our other meat eaters didn't kill as much because it was so hard to catch what they ate. Our species got to the point where not only did a person not have to catch the animal he ate, he seldom even saw it alive. The normal time a *sapiens* female in the 20th century came into contact with a part of an animal, it was in a supermarket, wrapped in plastic.

"It is true we were dominant over the large animals, wiping out a great many, but there were still a number of other animals that continued to compete with us quite successfully. Those were creatures we did not know how to get rid of, such as the housefly, mosquito, mouse, and rat. Our vaunted technology enabled us to dominate the big ones, but to the very end we were not sure we would win the battle against the small, prolific ones. The microscopic creatures, bacteria and virus, also continued to do quite well."

"And you claim that your main technique for domination of other creatures and the exploitation of your environment was through the use of these implements and machines?"

"Yes the things we learned how to make. But this also included modifications of nature which were not actual objects, such as irrigation systems. In a sense any of the techniques for modifying nature is what we called our technology. For instance one major achievement of mankind ensured a more reliable food supply which, of course, is the primary need for survival. It was called the domestication of plants and animals, and consisted in modifying the life pattern of wild creatures so they could be used more efficiently by humankind, mostly for food. I already mentioned how we learned to castrate animals so they would be less vicious and produce more tender meat. But domestication involved many other new practices so that in the end, the domestic animal and plant was mostly dependent on human care for survival, even though the creature was eaten by its caretaker."

By this time I recognized the yellow color code with the overlaid word "comment" or "question." I mused that it was the electronic equivalent of the raised hand in the classroom. So I paused.

"Yes, I understand the domestication idea. We on our world, as well as people on many other planets, have done the same, though with many different kinds of creatures. It seems inevitable that when a species achieves the capacity to do so, that it takes over others with lesser ability and uses them for its own benefit."

I sighed. "Ah yes. We humans became killers on a massive scale, like no other creatures on the planet. We came to the point of killing so extensively and processing the bodies that most of us never questioned its propriety. I know this well and it bothered me so much that some forty years ago I became a vegetarian. The great majority of mankind, however, kept eating the flesh of domesticated animals."

"But anyway you claim this domestication of creatures had significant consequences for your species, right?"

"Yes, in particular it allowed for many more people to be fed, which led to a great increase in population. This laid the groundwork for our greatest social invention, the city. Men could produce and store grain in vast quantities. Living in cities prevailed after about 5,000 years ago."

"So, I take it that your domesticators and urbanites eventually became the inheritors of what you call mankind's culture?"

"Yes, the greater productivity, the greater numbers of people and improved technology, especially weaponry, of domesticators and urbanites made them the inevitable rulers. There were still some hunters and gatherers in 1492, but they were already on the wane in most of the world. Which is how the world was when Euroman burst upon the scene."

Thus, Euroman took along on his voyages a relatively advanced technology. Though primitive by today's standards, these techniques were generally superior to those of the colored people who became aware of them even before they saw the pale-skinned bipeds themselves. Perhaps the most surprising sight was the ships cruising offshore. The vessels of Hernan Cortes were seen by the Aztecs first at Vera Cruz. The vessels were so large, the Aztecs thought they were houses on boats. Within two years, the empire was totally destroyed by the conquistadors who started the Spanish empire of New Spain, much as the world has been taken over by the Atierrans.

These and other sailing ships were difficult for many native people to comprehend at first. They were much more complex than the local equivalents, a harbinger of the technological marvels to come. The ships of Euroman were soon seen in one place after another. The Africans saw Portuguese ships cruising their coasts, and soon after the Indians saw Spanish, Portuguese, and English ships, as did the peoples of the Pacific. Even in the civilized nations of the Far East, the ships were a strange sight at first. And it was not long before it became apparent that the armed cruiser was a real threat, which started with an armed landing party and ended with extensive shelling of the cities of China in the era of "gunboat diplomacy." The Japanese were "opened up" in the 19th century by Commodore Perry when he steamed into Yokohama Harbor to issue an ultimatum for free trade from the U.S. Little did he know that he had opened the biggest Pandora's box in history.

The ships were, of course, basic to Euroman's expansion. They were able to explore the open oceans for only the second time that we know of in human history. The Polynesians had been the first open seas sailors, using their outrigger canoes, star maps, and knowledge of currents and waves for navigation. But apart from these craft, their technology was relatively simple.

The adventurers from Europe had borrowed some technological devices which gave them a distinct edge. For one thing, they used the compass, an invention of the Chinese, which enabled them to keep on course even when they couldn't see the heavenly bodies. The Chinese had never exploited the device to its full potential, trying to keep within sight of the coasts, as sailors had been doing for 8,000 years. Once the European seamen landed at the first unknown place, and lost their fear of the open ocean, they abandoned the coasts forever. Before the compass inventors knew what had happened, Euromen were in their harbors, as well as in all other sea lanes of the world.

It was one thing to cruise off the coast of a new territory, studying the shore, but it was quite another when the first landing parties went ashore. Euroman soon learned that except for a few small, isolated islands, these new places were inhabited. There were natives everywhere. Generally they were friendly at first, though this state of affairs usually did not last long. The greed of Euroman was quickly demonstrated. First the European ships were after provisions, especially after a long voyage. After that they turned their attention to precious stones and metals, spices, and other easily transportable valuables. On one shore after another the natives got the message. Further, as is the way with natives, they tried to hang on to what they had. But they also quickly learned that these visitors not only had marvelous devices for crossing the waters, they also had many devices and creatures for raising hell on land.

Probably the most significant objects the Euros brought ashore were those made of metal. Metallurgy went back about 5,000 years. Before that men relied primarily on rocks and wood for tools and weapons. But the chipped and ground stone tools still being used by tribals when Euroman landed rapidly went the way of the Dodo. An iron pot was much more efficient than a low-fired pottery vessel. The Euro-military (sailors and soldiers were basically interchangeable in those days) were also armed with weapons, armor, and trappings of iron. And they had already developed the most deadly killing device since the bow and arrow, the firearm.

It has been characteristic of *H. sapiens* to view his current accomplishments as the most stupendous. Toward the end of the 20th century we raised a hue and cry about the horrors of nuclear weapons. And while I could readily see the devastation possible, it was still hard for me as an anthropologist to ignore the terrors brought on by previous technological advances. A stone tipped spear seems unthreatening today, yet it was the weapon that started men on the road to dominance over the other large species. It is thought by some archeologists that the disappearance of many cold-weather mammals at the end of the Ice Age, the hairy mammoth, mastodon, rhino, wild horse, cave bear and many others could be attributed to the man armed with a spear. He trapped them in caves or swamps and ran them over precipices and into cul-de-sacs, ultimately spearing them. It does not take much imagination to realize what happened to many of the small tribes and to individuals who did not have the new weapon. The same undoubtedly occurred when the bow and arrow came along. What rag-tag primitive group of spear welders could hold its own against those armed with bows and arrows, with their greater accuracy and power? And then came metallurgy whereby spears, swords, and armor could be made of a relatively indestructible material. The Aztecs had pretty much the same weapons as the Spaniards except they were not made of metal. This undoubtedly contributed to their quick defeat.

Moreover, the Spaniards and all other Euromen after them, had that marvellous killing machine, the firearm. As usual, the basic component, gunpowder, was not invented by Euroman, but by the Chinese. However, it took Euroman to grasp the great killing value of gunpowder by using it within a metal tube to propel a missile. Euroman really took to this weapon and refined it through the years.

It is interesting to note that the particular Euromen who brought gunsmanship to its ultimate perfection were the Americans. They conquered the West with their version and armed their urban lawless, then enshrined the device in their media. The American attachment to the gun largely prevented urbanized governments of the late 20th century from keeping the killing machine out of private hands. Moreover, the entertainment industry relied on two primary means for solving dilemmas, true romance and shooting each other. If both these story devices had been taken from the American television industry, it might have collapsed. Clint Eastwood and Sam Peckinpah were as American as the hamburger.

In any event, in the 480 years between Columbus and Eastwood throughout the world European soldiers-sailors used this technology to take over territory. It was not really necessary to shoot all the natives in order to get their land and other wealth, although some Euromen did go that far, figuring it was the cleanness solution. The Australian Euromen in their elimination of the Tasmanians, and the Argentineans in wiping out their Plains Indians come to mind. However, it was usually necessary only to set an example. The Nazis, the most recent bad guys of the world, and efficient gun wielders in their own right, learned this lesson well in their drive to dominate other Euromen. If the locals did not cooperate, Nazi gunmen would shoot a few in a conspicuous place, then see to it that the message got around. The Euro-invaders in their takeover of native territories usually used the same procedure.

This was sometimes called the "demonstration effect." Some practices had higher demonstration effect than others. It was easy enough to show how effective a firearm was compared to most native weaponry. One could shoot a few natives armed with spears or bow and arrows and there would be little further resistance. The demonstration effect was high. In contrast, the demonstration effect of a religious practice was very low. How could one quickly prove that cutting a chicken's throat and sprinkling the blood on a shrine would cause one's brother to regain his vision; or for that matter, how could one prove that doing the stations of the cross fifty times would cure one's husband's cancer? Since religious practices generally had a low demonstration effect, people were expected to accept them on faith rather than by proof. And also for this reason, all peoples of the world wanted guns once they found out about them, while many people resisted Christianity and other converting religions, no matter how vehemently the Euro-missionaries exhorted them. That Christianity was spread as widely as it was seems primarily due to the indefatigable persistence of the true believers; while the spread of the firearm needed nothing more than the shooting of a few people.

Unfortunately for them, getting guns hardly helped tribal peoples and many others because by the time they got the first guns, Euroman had already overrun their territory, frequently by the use of better guns. The Euro-American inventors in Connecticut and Massachusetts were busy improving firearms. On the other hand, the great civilizations, those which had gained immunity from the White Plague, became capable of producing their own arsenals, after which European superiority ended.

The later explosive weapons also were quickly counterbalanced by the increasing technological capability of the colored peoples. Probably the last great edge of Euroman in weaponry was the atomic bomb, used as might have been expected, against a non-European culture which did not have the weapon. But as was widely known, this superiority lasted only a few years, after which many cultures that had survived the onslaught of Euroman made the bomb on their own. It is true that at Takeover, the major arsenals of nuclear weaponry were still in the hands of two Euro-powers, America and Russia, but many other nations also, particularly in South and East Asia, either had them or could easily make them.

So, the colored man of the world first saw oceangoing ships, from which pale-skinned men emerged bearing metal weaponry, particularly firearms, after which the local situation rapidly deteriorated. Natives in the interior did not see the ships though they frequently heard of them. Instead, their usual first view of the intruder was on other transportation devices. Frequently Euroman appeared either astride a strange animal or in a vehicle moving on revolving log cross sections. The animal, of course, was the horse, and the vehicle was some variation of a wagon. Again, what colored man saw in particular depended on where the landing took place since the domestication of the horse is believed to have taken place in Central Asia and the wheel in the Middle East. But still these two items were thoroughly integrated into Euro-culture when the palefaces began their great foreign adventure. Archeologists have revealed that the Middle

East was the primary world center for the invention of most important implements, plants, and animals, almost all of which were well established in Euro-societies by 1492.

The horse had long been used as a weapon of war. As a matter of fact, mounted horsemen constituted the last great threat to European cultural survival before Takeover. The men who assaulted them most were the fierce Central Asian nomads who were finally subdued by mounted Euro-horsemen bearing superior firepower. Even so, the nomads, astride their tough little horses, had by then fought their way to the gates of Vienna as well as into Peking and Delhi, the composite bow their main weapon. They were ultimately pacified by mounted Cossacks in the north and culturally absorbed in India and China. They managed to survive as ethnic independents in two countries, Hungary and Turkey, though eventually even in those places they were powerfully acculturated and interbred by nearby Euroman (woman). They did manage to keep their central Asian languages.

Another late threat to the continuance of European culture was Islam, also spread largely by mounted horsemen. As horse breeders well know, the greatest race horses ever developed were the Arabians, which were used effectively during the Islamic expansion. Starting from Arabia, they reached the Atlantic and Pacific in short order, conquering as they went. They got as far north as France, but were then progressively driven back by mounted Christian knights. The Spanish horsemen who attacked the empires of the Aztec, Inca, and other New World centers were descendants of those who drove the Moors out of Spain.

The horse gave the Plains Indians of North and South America a brief respite from destruction. The Comanche, Ute, Puelche, and Tehuelche were not stupid, and when they saw the incoming Spanish expeditions outfitted with horses, they began trading and stealing the animals and their trappings. By the 18th century the Plains were full of mounted tribes, hunting buffalo in North America and the guanaco (a small relative of the camel) in South America. However, they suffered the same fate as the mounted horsemen of Asia. The Euro-Americans, having taken over the eastern coastal areas by then, were steadily pushing south and west, using horses, but also bearing ever-improved firearms. Their technology was on a roll; they were rapidly following in the footsteps of the founders in Europe by going industrial. But even so, there were still the Indians, hunting buffalo and contesting Euroman's God-given right to take over and spread civilization.

Years ago when I first got into anthropology, I read a book of history, "Comanche, Barrier to South Plains Settlement". The historian, who was from Texas, claimed that Comanche raids had held up settlement of Texas by Anglo-

Americans for 100 years. He assumed that Anglo-Americans had that God-given right and barely discussed the fact that the Mexicans also had been there for quite a while. That it should be U.S. territory had already been proclaimed in the concept of "manifest destiny." Mounted horsemen, bearing the weaponry of New England, was the answer. The Comanche, as well as all the other North American Indians, were defeated and most of them banished to Oklahoma Indian Territory. The gaucho and local cavalry, also with better guns, finished off the Plains Indians in South America. They applied the final solution by killing them all.

Is the horse a part of a culture's technology? If my definition of technology as being the techniques for exploiting one's environment is accepted, then I think all domestic animals are such. The horse of course had other uses, being the primary means of land transport for Euroman, both astride the animal and using it to pull something, a wagon or a plow. In America our Euro-ancestors first used the horse to pull their wagons containing basic supplies into the interior and across the continent, and later used the animal as the prime worker for agriculture.

Europeans had a number of other domestic animals not familiar to many natives. Cattle, sheep, goats, pigs, and chickens, most of them originally tamed in the Middle East, were all new in the Americas and eventually became integrated into the life-styles of the subdued peoples. These animals were already part of the local technology in most parts of Asia, but in most parts of Africa, Oceania, and in all of the New World, most Euro-animals were innovations, and when introduced into local societies, caused important changes.

During most of history and prehistory, man has relied primarily on human muscle. Even up until relatively recent times, impressive achievements were made through nothing more than human muscle power and relatively simple tools. It is believed that the great pyramids of Egypt, as well as the stone fortresses and pyramids of the Aztec, Maya, and Inca were all built by men using simple stone-cutting tools, ropes, levers, and inclines, plus human muscles.

A great leap forward for men, not for animals, took place when man learned how to harness some of his domesticated creatures, particularly oxen and horses. From then on, there was animal muscle power, at least for those with the appropriate beasts. The sod of the American prairies was "busted" by the moldboard plow pulled by teams of horses.

Also, men toyed with what we now call alternative sources of power, particularly those of water and wind. The windmill and water wheel harnessed small parts of air and water movement and gravitational pull. And, of course, the expansion of Euroman throughout the world was largely the result of one important wind device, the sail. It is not conceivable that Euroman could have crossed the seas if he'd had to depend on muscle power as did the ancient Romans. Those other transoceanic travellers, the Polynesians, also used the ship's sail.

When the first men left European shores, there already was intellectual ferment, and within 300 years a whole new revolution in power use began the industrial revolution. We called it a revolution even though it did not occur in a day or even a week or a month and people did not set up barricades on the streets. It got its name because it not only brought a change of great significance, it shook the world. And it was started by Euroman who came to realize that gaseous pressure was a great source of power. Steam and internal combustion engines were invented, resulting in all kinds of changes in manufacturing and transportation. And when fossil fuels, coal and oil, were discovered to be effective for these engines, there was no stopping Euro-technology. Other discoveries, particularly electricity, came soon and Euroman began turning out goods at an unprecedented rate, soon to be peddling them around the world. The era of international trade had begun.

How did this tremendous development affect the colored men of the world, almost all of whom were under European domination by then? In the first place, the colored peoples were inundated with the manufactured goods of Euroman, blankets from Hudson's Bay to the Indians of Canada, cotton cloth from Manchester to the villages of India, and candy to tribal and village people everywhere. There were many negative consequences for the economies of the darker-skinned peoples, the main one being the destruction of hand-manufacturing.

A second consequence was a grabbing of the world's natural resources by Euroman. The new technology needed raw materials like cotton, sugar, rubber, and lumber to make into things, and the Euro-manufacturers needed fuel to feed their insatiable machines. And though they had some to start with, these were either not enough or were used up quickly. Where to go for more? Where else but the newly dominated lands? So the mountains, valleys and forests of the colored man became the source of cheap natural resources which have been carted off at an ever-increasing rate for the last 200 years.

Later, of course, the situation changed greatly because the center of industrialization shifted from Europe and America to the Far East. While once the British hauled uncounted shiploads into the ports of Liverpool and London, in the later 20th century the super tankers and freighters were hauling the raw material and fuel to the harbors of Yokohama, Seoul, and Taipei.

But perhaps the most important consequence of industrialization was a result of the demonstration effect. The colored man had learned early that to survive in the predatory world of Euroman, one had to have firearms. But then in country after country, almost all still dominated by Euroman, another message became clear: Industrialize or Perish! As fast as they could, without stirring up the wrath of the still-powerful Euroman, the colored men started to copy the new techniques. We know very well now what this led to. The societies that learned best and fastest climbed rapidly to the rank of Euroman, and a number are now surpassing the old masters. We saw the initiator of the industrial revolution, England, become a third-rate economic power while Japan and the other newly industrialized nations of the Far East became the manufacturing and trading giants.

There were other consequences of industrialization which shaped the world we lived in at Takeover. Probably the most over-riding one was the shift from the handmade item to the machine-made one. A power machine was really not very useful unless one could make numerous copies of the same item. So standardization and ultimately the assembly line became the way to make things. It is ironic that the specialists in standardization were the Euro-Americans who had revered individualism more than any others. However, the American worker insisted on his individual rights even while making a mass-produced product, and production suffered. He tightened the nut he was assigned to tighten with little or no interest in the product being made. And when the product fell apart or had to be recalled, he did not feel it was his fault. And as usual, a cultural rationalization evolved to explain the anomaly that the mass production, assembly line method automatically alienated workers. Since they weren't responsible for the whole unit, they only took responsibility for their own nut, and the whole product held no interest for them.

Unfortunately, Euro-Americans came to realize that their explanation was wrong. One could mass-produce goods and still come out with high quality. And one did not need to alienate workers. In fact, one could instill into the factory worker the same commitment to quality of the old-time hand-craftsman. But to do that, the worker had to be taught to be responsible for the whole product. The individual workers made up a team, whether it was a Taiwanese family or a Japanese company. And so, surprisingly, the spirit of the handcraftsman lived on in the member of the group and production leadership passed to the East.

6

The Symbolic Communicator

I got restless after smoking a bowlful of the slightly aromatic tobacco I had found in the market. I had never been one for super aromas, but a slight sweetness seemed appropriate. I studied my surroundings without noting anything particularly different except the chalkboard. I walked over and picked up the chalk. Evidently I had no particular intention when I began since I put nothing on the board but my name and a fantasy class:

Peter Hermann

Anthropology 1234

I studied it for a moment, then added, "The Last Anthropology Class." I remembered a novel I had read many years before in which the author kept referring to "final commencement." He, too, had been a college professor and used academic jargon in referring to death. So there I was, also thinking in such terms, but in this case, death of the culture. I could have put simply "The Last Class" because the other subjects of *sapiens* were not going to be taught in the world of the Atierrans any more than was anthropology.

Anyway, I thought that these sessions would constitute the final anthropology class, since the Euro-era had ended. Whatever might continue of its culture, including its classes, would be strictly for the benefit of the Atierrans and for reasons which we survivors were unlikely to understand.

When this line of thought ended, I began to enter the mindset of a professor again, something which still happened to me almost automatically if I had a chalkboard and a piece of chalk. I wrote the word "thought." Then without thinking much about it I wrote [θɔt]. This spurred me on. I wrote "etiquette" then [ɛtikɛt] and then "the," followed by [ðə].

The screen light came on gradually and I looked up to see the words, followed by the sounds, "Hi Pete. Busy again with your new device?"

For a moment I thought she was going to say "toy." But not old Mar, she was a diplomatic interviewer. The Atierrans and their electronic minions must certainly have had more training in interviewing natives than we'd had as anthropologists. If we were any good at getting reliable information from our native informants, it was only because we had learned how while on the job.

I answered, "You will remember I told you that a true professor always had his chalkboard. I remember going down the hall after a class, covered with chalk dust."

"Well, what are you writing now, Pete? It looks different, several words in some other script as well as the standard spelling?"

"That's it, Mary. The other script is the International Phonetic Alphabet, a method for depicting the sounds of words. It was devised by linguists to transcribe native languages. They found that the traditional spelling systems of most Euro-languages were so bad that it was impossible to write down a native language with the same characters and be able to read it back."

"I see. And you are saying that the writing in the brackets is more true than the other, that is, [θɔt] is more accurate than "thought" and [ðə] is better than "the," no?"

"Right. You see the problem was that spoken languages changed constantly while there was a natural tendency to keep writing the same. There was usually this push-pull effect when it came to change. We ordinarily tried to keep things the same even though it was obvious they were always changing. In my own culture men used to look back nostalgically to "the good old days" though when they analyzed the past, the old conditions were rarely perceived as better than the current ones. Things were no different with the tribal people. Among them one of the most common explanations for a particular custom was 'It's always been that way.' Anyone who knew their history, knew that was not so. It merely reflected humankind's desires. People were usually reluctant to change, though when it happened, few wanted to go back. This was even true of those who suffered disastrous changes like the tribals when they were overrun by Euroman.

Though they may not have wanted to lose their independence, the Indians, for example, had little interest in bows and arrows once they got guns or in carrying burdens on foot once they had pick-up trucks.

"The fact was that in most cultures, customs existed in the same form for only a brief time, though certainly some changed more rapidly than others. This was also true of language, but more in the spoken than written form. Thus the written form would get out of sync with the spoken one."

"Interesting, especially the ideas about change. But what does this have to do with today's interview?"

"Well, because it is high time to bring in language and this is a part of culture where though change was constant, many thought of it having a permanent ideal form."

"Okay, I'm listening, Pete." So I continued, "To begin with, language is basic to everything else in a culture. As soon as anything new is learned, one has to have new words to describe it and to pass the information on to the next generation. For instance, when Euroman overran the Indian cultures, he took over the words for the new vegetables and fruits, potatoes, tomatoes, avocados, maize, etc. In the same way and during my own lifetime a new system of calculation and information storage came to fruition with a whole new vocabulary, computer, software, hardware, floppy disc, hard drive, etc.

Apart from this very important function, language was the main means of holding a society together. Cooperation and the creation of histories, real and mythological, were made possible. One cannot imagine a religion without language.

"It is no surprise then that we characterize the species as a tool and language using animal, and especially language. After all, a number of other species used tools, but none had complex language systems consisting of thousands of units of symbolic meaning."

"Sounds reasonable. But how is your chalk business tied in?"

"Good question. As usual, I took something from my own experience to illustrate the significance of the topic. And you must know that I came through the back door of anthropology. In college I started out studying language, hoping to become a writer. Then I drifted into linguistics. All I had known about language before that was its literary aspect. I loved to read as a boy and even began scribbling at that time. But what made a language tick I had no idea. In my first linguistic class I was dumbfounded when the professor started writing words in phonetic script. Evidently I had already been uneasy about spelling even though I had been good at it. I used to win spelling bees, not knowing that the only reason these were so popular was that the English spelling system was so irregular. No matter what rules the schoolmarms taught, it was all a matter of memorization. So when I got the idea that a perfect writing system was one which had no silent letters and each written symbol had only one sound, it seemed marvelous. English, you know, has many alternative pronunciations and silent letters. That's why "thought" has only three characters in the phonetic script [θɔt]. The first one, the funny character that looks like a bisected lozenge, is not actually a combination of [t] and [h] but a separate sound which has no letter to describe it in the standard alphabet. The thing that looks like an open, flattened O merely means the sound as in "raw," also without a character in the English alphabet. And for [ð] the funny looking symbol that looks like a circle with a cross on top is the same sound as the "th" in thought except that the vocal cords are vibrated when it is articulated. The [ə] is simply another very common vowel in English for which there is no letter.

"You're really getting very technical, Pete. You know we don't need all the details of the sub-fields of your academic specialty."

I sighed. "Oh I know, Mary. I think I got a little carried away. And it's surprising in a way because I'm not the sort to get so concerned about the nitty-gritty. I've always thought of myself as a big picture man. As far as I was concerned, the accountants and bureaucrats and schoolmarms could work out the details. And I know from my teaching experience that the minutia of the science of linguistics drove a lot of students away. So, I'll try to curb my enthusiasm. Let it suffice that when I found out about the science of linguistics, I was hooked, and still am. It was the most meaningful analysis of language I had encountered."

"I suppose many other *sapiens* types have tried to explain the mechanics or other subtleties of speech?"

"Yes indeedy. Theories and explanations abound, as they should, I suppose. Nothing as important as language could be ignored, and the human way always was to come up with an explanation."

"An explanation?"

"Sure, some verbalization to indicate how logical an occurrence was."

"So what about an explanation for language from another source? Could you give me an example?"

"Sure could, Mar. For instance, the religious point of view. You may remember I said that religious believers frequently used supernatural explanations. Well, even by the time the Old Testament was written, it was clear that many mutually incomprehensible languages existed. So how to explain it. As usual, the religious types did not say 'I don't know.' They came up with an explanation which was duly written down and has lasted 2,000 years, to whit: Once upon a time mankind got uppity and decided to find his way to Heaven on his own. He began to build a great tower so he could march straight into the promised land without the difficult preliminaries the prophets had foreseen. Whereupon the Skymaster himself decided this wouldn't do and wished a language pox on the workers; whereupon everyone began speaking a different language, which resulted in royal confusion. Needless to say, that particular crew ceased to be very efficient and the tower of Babel, as it was known, was abandoned. But mankind was stuck with all those languages."

"I take it, Pete, that you hardly subscribe to this explanation?"

"You got it, Mar. I told you I got hooked quite a while ago on naturalistic explanations, and this is especially true concerning language. When I got into linguistics, I learned that the professionals would consider no other kind. So it didn't surprise me when I heard the linguist's explanation of why there were so many different languages. They had a branch that studied language diversity, called historical linguistics, and suffice it to say that by comparing old forms of languages with new ones, they learned that languages are always changing. Moreover, if a people went separate ways for 1,000 years or more, the end result would be different languages, like Italian, Spanish, Portuguese, and French, all derived from Latin. That was basically how 4-8,000 separate languages evolved not by the proclamation of a Hebrew skymaster."

"Pause" appeared briefly, then changed to "So it's linguistics today, eh Pete?"

"Actually it's language that we will discuss, but using the linguistics approach."

"Okay, go ahead. And first the developments before 1492, okay?"

"Sure. But again you must realize that most of what we knew about language before true writing was invented was conjectural. As usual, with customs which left no clear evidence behind, we had to make educated guesses for the early, early days. This was especially true of language. The best guess was that no creature with such progressively improving tools and increasing mastery of the environment could manage without a complex communication system, thus language.

"How do we know this? Well, when the anthropological specialist who searches for evidence of past cultures, the archeologist, got his first glimmerings, he recognized that such ancient tools followed patterns. In one area people made their tools out of the core of rocks, in other areas, out of the flakes. And as tools became more specialized, the types of tools kept becoming more particular. Usually the archeologist could tell by looking at it where a tool came from and roughly how old it was. This patterning through time and area told him that people were passing the same idea along, particularly from one generation to the next. These ideas ultimately became the basis of culture or man's way."

"To pass on a body of complex information, there had to be an efficient communication system, thus language. And though it has been noted that there were some differences in vocabulary, linguists were not able to find other qualitative differences. The Navajo language was just as good for teaching children how to exploit the environment as was English. English replaced Navajo as a result of political factors, not because it was a better language."

So when Euroman headed for foreign shores, he knew humans, no matter what other characteristics, were creatures who gabbled in languages.

And gabble they did as soon as the first shock wore off. And the colored men learned that Euroman was the same. Both groups talked to their own kind. Everybody had a language.

This is not to say that when Euroman landed on the beaches of the world that he believed the local languages, or any other customs, were as good as his own. In fact, the opposite was generally true. After all, not only was early Euroman not a linguist, the science of linguistics did not even exist in 1492. And Euroman, as the one in power, was a practicing ethnocentrist. Not only did he believe he was superior in technology (true), he also believed he was superior in all other kinds of ways, in morality, social customs, religious beliefs, food habits, clothing styles, art and language (all untrue).

But since all the natives had spoken languages, it behooved Euroman to try to use them. After all, like previous empire builders, Euroman knew quite well what an efficient tool language was for dominating and exploiting others. Apart from deceiving them, one could talk down to them, curse them, and ultimately cow them. Battle cries have been a part of intimidation in many cultures. It was claimed by some that the Oriental martial arts depended as much on intimidating the opponent verbally as using physical force. Furthermore, the best way to establish and keep a good slave or class system functioning was to develop a demeaning way of speech for those deemed lesser. Professor Higgins, an early socio-linguist, knew this well when he decided to change the social status of Eliza.

When writing emerged, man had a tool for super exploitation. He could keep records for taxing the workers. Writing in the hands of scribes was thought to be one of the primary means of keeping track of and using the peasants in the early civilizations. Language also was very useful for getting people to build pyramids, temples, roads, great walls, and cathedrals. The scribe never went away, of course. He merely changed cloaks. Well before Takeover he was known as the income tax agent.

But to continue, after storming up the beach with guns and/or swords at the ready, Euroman was faced with a universal human need to talk to the natives. What to do? Sign language was undoubtedly used in most first encounters. One could manage some basics that way, point to some food or water or pantomime eating or drinking and the natives might hand some over. Euroman could even point at some gold ornaments and pantomime interest, hoping the natives would take him to some. Australian gold prospectors who went into the interior of New Guinea in the 1930s did remain mostly at this communication level and managed. But in general, when Euroman got down to serious long-term business like taking over the natives' land or enslaving them or settling a peace treaty on his terms, such primitive procedures would hardly do. The Australian police officers and administrators who followed the prospectors and set up the new government (which came to be known as Papua), used the more efficient means, translation, at least at first. As soon as they were available, Euro-colonizers used individuals who knew both the Euro-language and that of the natives. The translator sometimes had to know several languages plus a lingua franca. In the takeover of the Aztec empire, Cortez found an Indian woman who quickly learned Spanish. He also made her his mistress. It was not unusual to use captured women in this way since they remembered their original language while also quickly learning that of their captors. This could be thought of as cross-cultural pillow talk. A famous Indian woman in North American history was Sacajawea who went along on the Lewis and Clark Expedition as a translator. Other translators were those who were well travelled and had picked up new languages. An early type was the ship-wrecked sailor who learned the language of the natives.

Dependence on the translator waned after the early stages of Euro-intrusion. A new phase began when some people on both sides learned the language of the other. One of the earliest Euromen who did this was the Christian missionary who quickly recognized the need if he was going to be able to convert the heathen. Of course, the bilingual cleric was also useful to the soldiers and administrators.

And others also learned the local tongues, primarily in order to get something from the natives. On the other hand, the trader quickly learned some of the local lingo to get the greatest advantage in furs for his beads, iron pots, firewater, and candy.

In more highly developed cultures, particularly in South and East Asia, many Euro-militarists and administrators made a point of learning the local languages. After all, these were written speech forms with long histories, spoken by hundreds of millions of people, as compared with the unwritten languages spoken by only a few hundred or thousand tribals.

On the native side, many learned the language of their conquerors, frequently as a matter of self-defense. After all, the trader was out to take whatever amount of beaver or sea otter furs he could get for his firewater or gew-gaws from the tribals, while elsewhere the military or administrator was trying to get land through treaties. And people at all cultural levels had to recognize that the Euros had technological superiority. The power of words thus became crucial. Knowing what the Euros were saying was one of the last defenses left to the native. Thus, many natives picked up some Euro-language. Even so, it is highly unlikely that many Indians got their beaver and sea otter skins' worth in Hudson's Bay blankets or Venetian beads, nor did the Chinese for their tea and the East Indians for their silks. The Scottish trader in North America and the English merchant in China and India were mean hagglers and were not to be outdone by their descendant from North America, the Yankee Trader. The other's language was one of the most valuable tools in the haggling that took place.

A particular problem arose wherever there was a great variety of languages, and especially where the Euro-language was not imposed absolutely from the beginning. In such places it was hardly worthwhile for the Euro-intruders to learn all the new languages. But they needed to communicate, primarily for trade or body snatching (slavery or indentureship). Of course, where the Euros decimated the native population early, language became irrelevant. This happened in the Caribbean, most of lowland South America, Australia, and Tasmania, as well as most of North America. Even so, a trade language evolved in the State of Washington and British Columbia, Chinook Jargon. But the main places where Creoles or Pidgins developed were in tribal Africa and the areas where black slaves were taken, mainly the Caribbean islands and the American South. One other region where a bonifide Pidgin developed was Melanesia, the large island region of the Southwest Pacific. In New Guinea and the surrounding islands there were hundreds, if not thousands, of tribes, each with a separate language. Moreover, though this was tribal country, it was fairly densely populated. Thus the British and their followers, the Euro-Australians, could not dispossess the natives as easily as they had in North America and Australia without massive carnage. So they fell back on the other main means of exploitation, body snatching. Just as the British and their descendants, the Americans, took to chattel slavery as a means of getting cheap labor, their brothers in Melanesia took to a form of semi-permanent bondage (indentured labor) which they graphically named "blackbirding." Those early marauding British could hardly be accused of being stodgy in their exploitation of the colored man. Not only were they inventive in the how, they came up with colorful terms to describe their depredations.

Anyway, when the Euros got involved in administering, carrying off bodies, and white-washing souls (converting heathens to Christianity) in the two big tribal areas of Melanesia and Africa, they found the complex language situation too difficult to cope with. So they simply talked to the natives in a simplified version of their own language.

One absolute need for learning a new language was to get some use out of it as soon as possible. If a modern North American wants a meal in Mexico, he uses what he has available. He leaves out many non-essentials, articles, adjectives, verbal variations, duplications, and sticks to the fewest possible forms of speech. Needless to say, he will not be concerned with keeping number, gender, or verbal form in consonance with the other words. If he can let the waiter know the basic name of the foods he wants, he will be happy. A child does the same, of course, when it first learns its own language. However, the infantile effort is corrected, first by the family and then by school teachers. And when adult immigrants learn a new language, they also are corrected. Old people usually do not learn to speak a new language without an accent, but they rarely perpetuate their erroneous versions on their sons and daughters who learn the standard version from their peers.

But when Pidgins developed, and I'll stick to Melanesian, the inaccurate forms both in sound and grammar were frozen early. But the newly emerging language served such a vital communication need in the extremely fragmented linguistic area of Melanesia, the locals got used to it in its early form. And before a school system could be set up, a language that was derived from English, but incomprehensible to standard English speakers, was in place. Later, proper schools were established and standard English taught. In the meantime, back in the hinterlands, Pidgin became more and more entrenched. Thus, when the state of Papua was set up according to the Euro-plan, most people were speaking at least three languages their tribal vernacular at home, Pidgin for intertribal purposes, and standard English for government or city affairs. The official languages for the new state were Pidgin and English. Even the Euro-missionaries buckled under, learning Pidgin as the language for soul snatching.

The dialectic form of English spoken by the African slaves and their descendants did not move far from standard English and thus could be considered dialects rather than separate languages. A standard English speaker could understand most Trinidadian or Barbadian English even if it did sound "funny." I know this from direct field experience in Trinidad where it took only a couple of weeks to understand almost everything said. The same was true of "black" English in the U.S., despite some bizarre Hollywood versions. Many expressions sounded different but a standard English speaker could understand most of it.

When the Euros hit the beaches of the world, they all had writing systems for their languages. The invention of writing seems to have taken place about 5,000 years before in the Middle East and East Asia. The Aztecs also had started putting their spoken form into characters.

And whatever else writing was, it was a way of freezing speech, of keeping it in a relatively unchanged form. In societies where no writing existed, it was difficult to keep good records, including histories. Though some peoples learned to tell verbal histories well, this was always less reliable than the written form. And when one considered *sapiens'* natural penchant to distort history, even the written ones were usually suspect. But at least they would not change simply from being told.

There were other advantages to writing, an important one being the ability to send messages over long distances. This was especially useful in developing war strategies, and undoubtedly helped the Euros a great deal in conquering tribal peoples. Moreover, once the land was taken, detailed descriptions could be written to send back to headquarters. It was no accident that the "father" of the United States was a surveyor. Surveyors were in the business of measuring and describing the land which had been taken away from the Indians. And their measurements went back to headquarters ASAP, ordinarily on documents, written pieces of paper.

The printing press, brought to full development in 16th century Europe, enabled Euros to make many copies of their reports and other documents and to disseminate them widely.

The great majority of languages were not written. When the Euros came, none of the tribal peoples could read their own speech. In fact, the usually accepted definition of a tribe was a society which lacked writing. So they watched in amazement as the Euros proceeded to put chicken tracks on paper and later read it back. And though this facility seemed to give the Euros great power, to most tribals it seemed to be something intrinsic to Euro-languages and not their own.

In the meantime, more and more tribals were learning to speak Euro-languages. Despite the belief of many school teachers, there was absolutely no necessity to know the writing system to learn a new language adequately. Even so, non-Euros had been exposed to the writing of the palefaces. And inevitably some tribal individuals decided to do the same for their own languages. They would usually modify the Euro-version of writing for their own speech forms. Then, they believed, they too would be able to send unchanging messages and set themselves free from the inaccuracies of memory.

The newly invented writing system I knew most about was the Cherokee alphabet, or more properly syllabary. The Cherokee Indians of the American southeast were a remarkable people in many respects. After being overrun by Euro-militia, they got the message and decided that the only way to go was as Euros. They would have to adopt Euro-American practices in order to survive. And they went far, replacing their old housing type with the log cabins of the newly entrenched Euros, trying to practice Euro-technology, including feeding Indian corn to pigs and chickens instead of eating it directly as they had done before, adopting Euro-clothing styles, and even trying to set up a Euro-style government with a bicameral legislature, constitution, etc.

So it is hardly surprising that someone among them decided a writing system was in order. He was a half-breed Cherokee, Sequoya, who had received a smattering of Euro-schooling. He devised a writing system for the Cherokee language, utilizing what he then had available, an English writing book. And by manipulating the characters, he came up with a sound-based writing system, what linguists call a syllabary. This was the only case that I know about of a real writing system being invented by an insider.

One might imagine that by trying so hard to be Indian "whites," and even after one of their people had invented a writing system, that the Cherokees might have been treated better than "savage" Indians. It didn't work out that way. Most were dispossessed of their land anyway and put through one of the worst ordeals any Indians had to suffer. It was called "The Trail of Tears," a forced evacuation of their homeland to Indian Territory, in the process of which a high percentage died from poor food or the lack thereof, illness, and the arduousness of the trip. I suppose if people did die of broken hearts, the Cherokees would have had reason. But still, when the survivors settled in Oklahoma, they put their new writing system into use. The only other positive thing that came from their pains was to have a species of tree named after the inventor of the writing system. As usual, the Euros took almost all the land. The few Cherokees who managed to hide out in their home territory in the Appalachians ended up as entertainers, doing pan-Indian dances for Euros in the Great Smoky Mountain National Park. Later on, of course, even the last Indian territory, Oklahoma, where most surviving Cherokees ended up, was opened for Euro-settlement.

I believe the Cherokee writing system was still in print in Oklahoma at the time of Takeover. This and the name of the California tree were the consolation prizes.

Many writing systems were devised for tribal languages by Euro-linguists. Of the various types of Euros who descended on native peoples, one, the religious cleric, was particularly interested in local languages. Unlike the administrators and military, the Christian missionaries perceived an advantage to learning and using the local tongues. Their primary goal, conversion of the heathen, had required the use of local languages ever since the earliest wave of proselytizing when missionaries went from the Mediterranean cities north to the lands of the then heathens, the Huns, Anglos, Russians, and other barbarians. In order to convert them, the missionaries believed they had to communicate. After all, the primary technique of Christians, and perhaps other clerics, for spreading the word was repetitious exhortation. It must have taken a lot of fancy preaching to bring the nature worshippers of northern Europe into the fold. Threats of damnation and promises of beatific eternal life could work only if the hairy heathens got a minimal understanding of what the preachers wanted them to do. And since the European savages had no more writing than the tribal savages of other parts of the world after 1492, the missionaries worked at it, mainly on the Germanic and Slavic languages. The earliest writing form of German, Gothic, was devised by a southern bishop called Ulfilas. And as might have been expected, the new writing systems were first used for making translations of the Bible and prayer books.

So when the Euros were spreading after 1492, a long-established tradition already existed. The new missionaries quickly learned to speak the new languages and then devised writing systems for many. And though their primary purpose was to use the native languages to promote their religion, once an adequate number had learned the new writing systems, they could be used for other purposes.

Thus, there are now writing systems for practically all the surviving native languages of North America, as well as Africa and other tribal areas where missionaries got busy before the native tongues died out. I knew missionaries from the Oklahoma Linguistic Institute who were studying the local languages of the mountain tribals in Laos in the 1960s. Whether they got around to devising new writing systems for their converts, I do not know. The problem was that another aspect of Euro-culture interfered, the war syndrome. All hell broke out in this area when the U.S.-Vietnamese war spilled over into Laos. And since the Vietnamese won the war, I suspect the American missionaries quickly became *persona non grata*. It was not as simple as working with tribal remnants of American Indians on reservations.

This new type of missionary became very good at language analysis, working alongside the anthropologist in using the scientific methods of descriptive linguistics. As a consequence, the writing systems they devised tended to be quite good. My first introduction to this field was through a handbook by missionary-linguists, Pike and Nida.

This is not to say that everything came up roses in language contact. There were contrary forces. One was that multitudinous native languages died out early because the speakers did. When the last Tasmanian in Australia and the last Yahgan in southern South America died, there could be no further continuance of either spoken or written Tasmanian or Yahgan. The small tribes were wiped out most quickly by the overwhelming force of the Euro-juggernaut. California, an area which had a large number of tribelets, lost many tribes along with the languages, when the Euro-Americans seeking gold came streaming in.

We have one graphic account of the last "wild man" who was the last speaker of such a dying language. This was Ishi, the last speaker of Yahi in northern California. Ishi, along with his few surviving fellow tribesmen, headed for the Sierra ridges in the late 19th century when it became apparent that the Euro-ranchers and prospectors would kill them all if they stayed around. The little band survived as tribespeople until 1911 watching the bustle of Euro-expansion in the valleys below. But when the last other Yahi disappeared, Ishi decided he had to join the dreaded Euros. He came down from the mountain top and was caught in a cattle pen in Oroville, California by amazed Euros.

The problem was that no Euros knew how to talk to him, though it was apparent that he had a language. Accounts of his existence made the newspaper and ultimately reached the eyes of anthropological linguists at Berkeley. Through the new science of linguistics, Professors Waterman and Kroeber established communication and ultimately worked out a way to fit Ishi into university life. The last wild Indian seemed to have had a fairly happy time in his remaining years. He learned English while working as an informant for his anthropological benefactors and as a janitor in the university museum to fulfill the bureaucratic requirements. But he had only five years to live as a Euro, dying from tuberculosis as so many of his brothers were doing on the reservations.

It was a different story with the great civilizations of Asia. They had been literate for several thousand years when the Euros first appeared. In fact, the Middle Easterners were the original inventors of writing. All had fully developed written literatures. Thus, though the Euros could foist their own languages on the newly colonized, there was little to be done to continue or improve the existing languages of civilized peoples, much less devise writing systems for them. As a consequence, far more Euros learned Hindi, Chinese (Mandarin), and Arabic than such small languages as Modoc, Pomo, and Potawatomi. To be successful with civilized people as a missionary or trader, it was more efficient to deal in their own languages: Japanese, Chinese, Urdu, Hindi or Arabic.

But wherever the Euro-intruders set up bona fide colonies, they introduced their own languages for administrative and later educational purposes. Thus, the British introduced English in their new colonies in Africa while the French did the same in theirs. English became the language of administration in India while in the Philippines, Spanish served that function in the early period until English replaced it when the Americans took over. Dutch was the official language of the colony of Indonesia while French was introduced into what was then called Indo-China.

Something special took place during the colonization of India which was to greatly influence the future study of languages. Some Britishers with a scholarly bent discovered that the ancient language, Sanscrit, was related to most of the early languages of Europe, as well as to those of Persia and Afghanistan. Their comparison of cognate words was the beginning of the science of historical linguistics which traced the relationships of one language to another. The new linguists came up with a super-family of peoples, Indo-European, as widely apart geographically as Norwegians and Bengalis, all of whom were thought to have left their original homeland in Russia some 5,000 years before. Peoples who were separated geographically tended to develop mutually unintelligible languages in 1,000 years or so. The daughter languages of Proto-Indo-European then became Sanscrit, Old Persian, Greek, Latin, and Old German.

Another consequence of this finding was that the ancient Hindus had long been interested in language study on their own. A great Hindu scholar, Panini, had done a masterful analytic study of the ancient language which helped stimulate European scholars to do the same with their own. Thus this laid down the other base for linguistics, the analytic study of languages.

It might be worthwhile to take a little jaunt around the world, to see what languages would have continued into the future if Takeover had not occurred. The last cultural baggage left behind by great empires was frequently their languages. When other powerful political systems collapsed, like the Persian, Greek, and Roman, their languages continued on for several hundred years afterwards. Like a fair number of my contemporaries, I studied Latin in high school, even though by then it had not been a viable spoken language for over a thousand years.

First, were the Spaniards, the original conquerors. Everywhere they became established, they introduced the language of Castille. In the two empires they conquered, Aztec and Inca, Spanish was first used alongside the local tongue, serving as the language of control, while as usual the local speech form continued to be used by the Indians among themselves, especially in the households. But the prestige language was Spanish from the beginning. If one wanted to be fully accepted in the new society, one had to speak Spanish. Latin American anthropologists have claimed that discrimination since the conquest was based not so much on physical characteristics as it was in North America, but on whether individuals practiced Spanish customs. An Indian was not so much a person with high cheek bones and straight black hair but someone who spoke an Indian language and followed some other Indian customs.

Inexorably, the native languages began to die out. This process was almost complete in Mexico by Takeover, though in Guatemala many Indians continued to speak Mayan, while in what had been the Inca Empire, Quechua and Aymara continued to be spoken.

Where there were no true empires, Spanish replaced the local languages much more rapidly. Thus the other countries of Latin America, except Brazil, were almost entirely Spanish speaking, the Brazilians speaking the language of their Euro-intruders, the Portuguese.

The Spanish had more territory where they had established their language, notably in North America, the Caribbean Islands, and the Philippines. But the North Americans grabbed most of this, at which time they replaced Castillean with the language of Shakespeare. The large islands of Cuba, Santo Domingo, and Puerto Rico were the only ones which continued to be Spanish-speaking in the Caribbean, though most people in Puerto Rico also learned to speak English. In the Philippines, except for a few words, phrases and names, the only people who still used Spanish were the elderly. But still, because of the original conquests, the second most important language of the Americas continued to be Spanish.

The other early New World Euro-language was Portuguese. The rulers of Spain and Portugal had divided that great plum, South America, between them when the New World was first discovered, the eastern territory going to Portugal, the rest to Spain. Thus the little Euro-enclave of Portugal got the enormous colonial fruit of Brazil. There were no Indian empires there, so the Portuguese progressively replaced the Indians and introduced their own people, language, and customs. By Takeover this vast expanse was 95% Portuguese-speaking and the few Indian languages which remained were of interest only to anthropological linguists and missionaries.

The super-colonials of the world were the British and their descendants in America, Australia, New Zealand, and South Africa. They established the English language everywhere, and in all the small tribes it progressively replaced the local speech forms. Thus, only a small percentage of North American Indians, Maori, Hawaiians, and Australian aboriginals continued to speak the old languages.

One of the interesting offshoots of English in tribal areas were the Pidgins or Creoles, simplified forms that served as lingua francas between tribes and were used by the English masters to control the dark-skinned slaves or indentured workers. In most parts of the world these became comprehensible dialects. The one place where a genuine new language evolved, basically incomprehensible to standard English speakers, was in Melanesia. It became one of the two national languages of the new state, Papua, English being the other. There was no native language spoken by a sizeable majority of the Melanesians.

British Africa went a slightly different route. The new colonies were put together for the convenience of the colonials. Thus, many tribal and ethnic groups were combined in one national package, there being no dominant language group. Nigeria contained a combination of languages Ibo, Ibibio, Ife, Yoruba, Hausa, Fulani, and a dozen or more smaller ones. When independence came, it was very difficult to get an agreement on a national language, so English was made the official tongue. The same process took place in other English African colonies and in several new countries of Southeast Asia. In these places the language of government, big business, and the educated elite was English while the natives spoke their own lingo at home and in the marketplace. Most of the village folk spoke only their own language.

One of the unexpected side effects of the imposition of a Euro language on tribals was the creation of a kind of unity that had not existed before. The tribals of the world had limited knowledge of similar others before the Euros came, usually those no farther than up the valley or across the mountain. Further, they usually did not speak each others' languages. So when a new language, English for example, was foisted on them, the tribals suddenly had entirely new possibilities for communication. The Indians of North America, especially those in the West who survived as independents longer than those in the East, are a good case in point.

The social situation was particularly chaotic among the Plains Indians during the last three hundred years of their freedom. When horses became available through trade or theft from the Spaniards in the 16th century, the Indians changed their life-style to a great extent, becoming mounted buffalo hunters who moved rapidly into the region where the shaggy herbivores thrived. The Sioux and Cheyenne, for instance, went from Minnesota to the Dakotas and Montana while the Utes and Comanches went southeast to Colorado, New Mexico, and Texas. All these people spoke their own languages.

One of the earliest partial solutions was the development of a sign language which concentrated on ideas rather than sounds. Because it was not tied to any specific language, it could be learned relatively easily, even by Euroman. It struck enough of a chord in the Euro-mentality to become part of the Indian lore of the Boy Scouts of America, thus continuing long after most other Indian customs had died out. One could earn a merit badge in the Euro-boys' club by learning some Indian signs.

Sign language was of little use to Indians once they were put on reservations. However, more and more "redmen" learned English out of necessity, mainly to argue for the rights the Euros had "given" them after taking away their right to live as independents. Also, knowing the Euro language was one of the few ways the Indian could avoid being cheated too badly at the trading posts. Then too, the Euros began to send the young people to Indian schools, frequently by force. The children were punished for speaking their Indian languages. Before long so many Indians knew English, they had a common language between tribes. The new language began to serve as a force for unification. Thus, a Navaho could talk to a Winnebago, both referring to themselves as Indians, an idea also introduced by the Euros. Before the reservation period Navahos did not even know that Winnebagos existed. A kind of pan-Indian culture evolved and new cults, mainly on how to deal with the Euro-threat, appeared and spread through the use of English. And in the era of civil rights in the U.S., from the mid-20th century on, some Indians organized to try to get a few rights back, invariably depending on the use of English.

In the tribal areas of Africa, much the same happened even though the native languages continued to be used in the villages and as the co-language among the educated of the cities.

India, Pakistan, and Sri Lanka also were polyglot combinations, so when independence came, they too could not agree on which local language to use nationally. It was obvious that English would remain as the language of government, higher education, and international affairs, but choosing a local language created much dissension. Largely because of such language differences, Pakistan broke into two, the eastern Bengali-speaking part becoming Bangladesh, the western part continuing with Urdu, both depending on English as the means of communication for government and international affairs. Both Sri Lanka and India faced continuing tension and terrorism for many years over language issues.

Even so, India and to a lesser extent the other countries, achieved a considerable amount of unity through the use of the imposed language. An educated Dravidian speaker from South India, who would not admit to knowledge of North Indian Hindi or Punjabi, would converse without hesitation with his fellow Indian in English.

As a consequence of British colonialism then, the adoption in so many places of English as the language of government, and because of its importance in industry, commerce, and higher education, it became the paramount world language. There was nothing exceptional about the tongue even if it was the vehicle of Shakespeare; and the writing system was abominable. But as a result of the exploits of those super-colonials, the British, and the industrial and commercial power, first of the English and then the Americans, English became the international language. Every country that wanted to succeed economically promoted the learning of English, and the only language the international tourist found absolutely necessary was English. And after the great shift of power took place in international commerce in the latter part of the 20th century, it was easy to see that the language of the super-colonials, the English, would remain long after other influences faded.

One other Euro-language that got to the semi-finals, but lost in the finals was French. The French did not enter the colonial sweepstakes with as much gusto as did the British, and simply did not become established in as many places. Even so, they established a respectable number of colonies in Africa, IndoChina, and a few other small places. In all instances, they introduced French as the language of government.

The French had gotten away with a reputation for high culture for about 200 years before the colonial era began. Euros who wanted to appear learned or sophisticated flaunted their French. It was quite the thing to know French in high society. This idea persisted well into the 20th century. When I was trying to shake my midwestern parochialism in the 30's and 40's, I memorized a few French quotes which were common in the literature I was then reading. However, after W.W. II using French for effect went out of style.

The idea that French gave one high status also was foisted on the newly colonized. Thus, the elite in the African colonies as well as in Vietnam, Cambodia, and Laos, learned French and flaunted it where they could. After W.W. II it became apparent that no matter how civilized their language, the French as colonizers could be as savage as the English or other colonials. Moreover, it soon became apparent that the language of the future would be English. Perhaps one of the last ditch efforts of the French to hang on to their reputation as the most cultured European country was to sponsor and finance a widespread educational effort in the ex-colonies of Africa and Asia. After the French military presence had become inconsequential, great numbers of school teachers who spoke only French were placed in the school systems of Laos, Cambodia, Vietnam, Algeria, and the central African ex-colonies to keep the French tongue alive. In most such places the new competition was English because this was the period when the U.S. leadership visualized their country as the new force of Western civilization. The U.S. Peace Corps was devoted largely to the spread of American English.

So despite the wishes of so many Frenchmen, their language was progressively replaced by English in country after country. And again this had nothing to do with the quality of the language. Despite the claims that French was the ideal language for romance and sophistication, from a linguist's viewpoint it was no better or worse than English, and its spelling system was just as abominable.

One other Euro-language, Dutch, became established for a while in a good-sized area. The Hollanders had one large colony in the heyday of landgrabbing, Indonesia, and true to the pattern, they introduced their own language for governmental affairs, international relations, and higher education. But after they lost a particularly bloody war with the Indonesians, they pulled out of their colonies as the French also were doing at the time. And as the old folks died off, Dutch declined in importance, again being replaced by English.

There were two other languages of great significance in the world at Takeover, Russian and Chinese. Russian was derived from the same language superfamily, Indo-European, as was English, French, and Spanish. The main difference in its spread was that the Russians stayed on their own land mass. They used their energy for colonizing Central Asia and Siberia where, of course, they did introduce Russian as the language of government, etc.

Actually, the Russians did have one brief fling away from their own continent when they invaded the Aleutian Islands, Alaska, and coastal North America. But this was historically late, and evidently the Russians did not really put their heart into it or were too busy elsewhere. And when they sold Alaska to the Euro-Americans for a pittance, the Russian influence and language rapidly withered in North America.

Chinese was spoken by more than a billion people and as usual as a consequence of national expansion and a high rate of reproduction. But even more than the Russians, the Chinese expanded mostly by land so the dominance of their national language, Mandarin, was mainly like a huge ink-blot spreading out from Beijing.

From the mid-19th century on the Japanese advanced greatly in the industrial-commercial circles of the world. And if their effort to dominate most of Asia by military force had been successful, Japanese probably would have become a great international language. But since they lost W.W.II and depended largely on using English for the war they were winning afterward— international trade, their language spread very little beyond their borders.

7

What's For Dinner

I bit deeply into the rosy, yellowish peach, savoring its juiciness, sucking in the excess liquid before it could run from the corners of my mouth. Also I pleasured in the fuzzy texture of the skin, knowing that there had been many persons in the pre-Takeover world who had not liked the fuzziness of a peach, who had in fact carefully peeled each one before eating it. I had not been one such, perhaps harking back to my boyhood in Indiana when all peaches had been fuzzy. I could not remember having eaten a nectarine in those days, the fuzzless peach developed in innovative California. Soon the peach was gone and I dropped the pit into a nearby waste container. I then studied the bowl of varied fruit, trying to decide whether to eat another. Finally I chose a pear-apple, a fruit that resembled a fat pear, though I knew the texture was more crisp and super-juicy while not being too sweet. But instead of biting into it, I got up and went to the small refrigerator that Mary had put in. I took out a bottle of Rosé d'Anjou, a fruity wine I had often taken in my lifetime. The small cabinet alongside contained glasses and an opener. Just as I was about to open the bottle, the now familiar sounds of computer start-up began, and before I knew it, there was the low background music, the Pastoral Symphony of Beethoven. The lights flickered and Mary began, "Hello Pete, how are you doing?"

For a moment I felt spied upon and guilty, but got over that feeling in a hurry. After all, there was no reason for me to doubt that I was under surveillance all or most of the time, and I had come here on my own. Further, I had been treated kindly so far. I answered, "I'm okay." After a pause I added, "I guess I should ask you the same question, except it doesn't seem appropriate. I had never thought of including a computer in my social niceties though I must admit that I had never dealt with any like you. Is it appropriate to inquire about your well-being?"

"It can be done, though I think primarily for the benefit of the animate respondent, you in this case. Animate beings, particularly the social ones, all seem to have developed communication procedures which almost always begin with some ritualized inquiry about one's state of being. Anyway, it's okay if you want to ask me about my health. And in any event, for now I'm all right, especially after the repair robot fixed that faulty circuit. We generally keep going for long periods. It's not often that we need repair."

I turned back to the bottle. "Do you mind if I have a glass?"

"No, that's why it's here. We are aware that you, like many other former middle- and upper-class Americans, were in the habit of taking the fermented juice of the grape while the working class tended to prefer the drink made of fermented barley and hops."

I removed the cork, poured a glass, and went back to one of the chairs, setting the wine on the coffee table. I waited a bit before Mary spoke again. "Is something wrong? Aren't you going to drink it?"

I felt uneasy again but answered anyway. "I feel a little uncomfortable. You know, we *sapiens*, especially Euro-people, usually drank our wine socially. I feel I should be offering you some, and yet I know that is not appropriate. Electronic devices do not consume food and drink. I suppose the problem is that I think of you almost as an animate being even though I am well aware that you are really just a very complex computer."

Mary's screen retained the word, "Cycling" for a minute or so before she spoke again. "I understand. This has happened before in cross-world investigations. But you are right. Even our creators and masters, the Atierrans, consume only for energy and growth, never for pleasure. So you go ahead and have your wine."

I sipped the cool, slightly sweet drink, admiring once again its beautiful, pink color. Mary came on again. "I see you also have a variety of your world's fruit. Where did you find those?"

"The same place I got the wine, in a well-stocked supermarket, the type we had in great numbers before Takeover. Now they are practically gone. This one was not too far from the little market where I bought the pipe."

"Oh yes, I think I know what place you mean. We left a traditional neighborhood in many cities, primarily so we could check out certain data. We had learned from our experience on other worlds that if we replaced everything, it was next to impossible to find out certain facts simply from books. For instance, how many history books discuss constipation as a problem, even though you Euro-people suffered from it extensively for 200-300 years. But during your takeovers, particularly by the Spanish, whole cities were replaced from which later knowledge was almost impossible to get. We understand that the replacement of Tenochtitlan by Mexico City and the replacement of other urban centers in Meso-America destroyed much data about those civilizations."

"No doubt about it. Later on, we tried to fill the gaps in knowledge by sending archeological expeditions to those places, but much of the information could never be recovered."

"Yes we know. And that is why we have a long-standing policy of not replacing what you call a culture until we are absolutely sure we know a lot about it. So, we left a few supermarkets in traditional neighborhoods."

I took another sip, then putting the glass down, took up the pear-apple again, preparing to cut it in half. Mary spoke, "That's quite a collection of fruit you have. You certainly seem to like variety."

"Yes I do. During my days of field research I was exposed to many different kinds, particularly in the tropics, and then when I settled down in southern California I had even more exposure. In fact, I joined an organization which existed solely to encourage the growth of tropical fruit plants in that corner of the United States, the Society of Tropical Fruits. And when the supermarkets put in special sections for exotic fruits, I quickly became a customer. It was one of the first places I would go when I went into a grocery store."

"And I take it that is where you got that collection."

"Right, and not a bad selection at that." I turned the pear-apple over slowly, feeling the grainy texture of the skin. "You know a collection like this is also an end product of Euroman's doings. It represents a social process we used to call diffusion. That is, things were spread within societies and from one society to another. It simply means that people borrowed from each other, in this case, new fruits. And with Euroman's extensive travels, it was inevitable that such borrowing would be widespread. Every time there was a new voyage, there was a likelihood of some new discoveries and some of the new items were brought back. This variety you see came from many different parts of the world, the pear-apple and Fuyu persimmon from the Far East, the sapote and cherimoya from the New World, the small red bananas probably from India, while the apple and peach were from the Middle East and Europe. Europe, however, at least the northern part, produced very new few varieties of any kind of edible."

"Hmm. I take it that you learned to like all these exotics."

"Yes, frankly I became a totally dedicated borrower. Consuming the same things over and over I found very boring, to the frequent consternation of my wife. In my later years I was always looking for something new to eat, though I continued eating some of the good things of my youth also. In a sense I belied my age in this respect. Generally, we *sapiens* were most innovative in our youth, and became more stodgy as we grew older."

I held up an apple and peach to show Mary the traditional fruit of my early years. "Interesting. Don't you want to eat some of the fruit you were starting on like the pear-apple?"

"Sure." I cut it in half and took a big bite, pleasuring in the juicy crunchiness.

"I've noticed that you seem to be able to bite well for a *sapiens* your age. Our investigators report that one of the main problems of your old age is the loss of teeth. Most of your older people get man-made teeth which come out of the mouth. Is that not true?"

"Yes, and I did also. I tried hard to avoid it because I had sad memories of my father's poor dentures and I was active in the dating game into my sixties. I even tried implants for a while, but when push came to shove I had the remnant teeth taken out and dentures put in. And I found it to be one of the most important medical changes I made. The dentures were fine and I only took them out briefly once a day to clean them. And dental care was much less than in the old days when I was trying to save my teeth – those I still had. With dentures I could eat anything I ate before and other people did not even know that I had false teeth. And for the first time in twenty years, I had no more dentists sticking novocaine needles into my gums, cleaning my teeth or the multitudinous other unpleasant procedures dentists had figured out 'to save the teeth.' And the monetary savings were enormous."

I probably chomped on the pear-apple harder than necessary. But then I knew I always took advantage of an opportunity to be histrionic.

Mary came on then with "Transition."

I had learned that this meant we were to go to the next topic, that we had dallied enough, even if I saw it as an introduction to the upcoming topic. So I waited. Then Mary said, "So, you seem to have a message of some kind now about food. I presume that was what the variety of fruits was about."

A sharp lady, even if electronic, I thought. But deviously I answered, "I don't know that I planned it exactly that way, though since the rise of the subconscious in our thinking, we have never been exactly sure what we have done deliberately and what has been the product of our un-admitted intentions. Anyway, I agree it is time to go on. And I do suggest that we take for the next topic food and drink. These were the basics for life on earth and it was inevitable that *sapiens* turned to his technology for their fulfillment. Okay?"

"Sure. But what were the basics before the rise of Euroman?"

"Well, it should be emphasized first that no part of technology was more important than the methods of procuring nutrients. Man could do without many kinds of social customs or beliefs but not food. A starving man was just that. He was not a Christian, Hindu, or atheist, he was simply dying.

"Even so, the basic techniques for most of the history of *sapiens* were little different from those of the other animals. He was a gatherer of wild products, both animal and vegetable, with a few minerals thrown in. He did better than his competitors as the millennia passed, mainly because his tools kept improving and his understanding of nature grew. By the end of the Ice Ages some 10,000 years ago, humans were already the most efficient hunters and gatherers on earth. And then they discovered in several places independently the possibilities of domestication. The human animal turned the animals and plants he was gathering into slaves whose lives he could control and whose bodies he could use, mainly as food. He became the domesticator and flourished. Such was Euroman when he overran the world."

Euroman had his own repertory of food items. The super grains were wheat, oats, rye, and barley while the animals were cattle, pigs, goats, sheep, chickens, ducks and geese. There were also a fair number of vegetables and fruits. Most had originated in the Middle East, but when Euroman left on the seas of the world, they were well integrated into his culture.

Most seeds had many advantages as human food, but they had one important disadvantage. Human dentition was not capable of breaking them up to make them easily digestible, to say nothing of what they could do to the teeth.

I had this human limitation dramatically illustrated several years ago when I took my wife on a trip to Taiwan. My son, who also became an anthropologist, was showing us around a village area at the end of the rice-growing season. The locals were drying their rice on flat areas around the temples. My wife knew nothing about rice at that time other than what was stated on a supermarket package. "So this is what natural rice looks like?" she said, picking up a small handful of the grain.

"Yup, that's it," I said.

Whereupon, "Oh, it hurts."

I stopped and turned to see her fingering her mouth. "What's wrong?" I asked.

"My tooth, I did something to my tooth." And she proffered a sliver of molar.

"What did you do?"

"I broke my tooth."

The tooth was repaired when we got back to the center of Euro-dentistry, the United States. But it clearly showed what would happen if humankind, with its puny teeth, had tried to break up grains with nothing more than the physical equipment of the evolutionary process. She had tried to crack only one.

Some peoples have been reported to have particularly strong teeth and jaws, the Eskimo for instance; but they chewed raw meat, not seeds. However, even that capability rapidly disappeared when refined Euro-foods, flour and sugar, were made available. The hard Eskimo teeth were just as susceptible to dental caries as the teeth of Western man.

Anyway, mankind solved this problem, as he did practically all others, through invention and learning. Methods have been devised throughout history to make up for biological deficiencies. That was the human way and the end product was what we called culture. So Euroman inherited from the peoples to the south quite a few methods of survival based on learning. One was the technique of cultivating grains, especially wheat; another was to grind it into flour and to make it rise through the use of leavening, then baking it.. Some 5,000-10,000 years ago some brilliant cook hit upon the idea of putting special bacteria into grain dough and letting it set overnight. The bacteria would produce little gas bubbles and raise the dough. Popped into the oven, the gaseous dough was fixed in the same shape and a loaf was born.

On a hill farm in southern Indiana, my grandmother followed an old process that could be traced back at least to the ancient Egyptians. She set aside one day a week for baking in the woodburning stove. She and her daughters would have made the dough the day before with wheat flour they'd had ground at the local mill and mixed with yeasty dough that was kept from week to week as a starter. For most of the day they kept popping the raised loaves into the oven, spreading them on shelves to cool afterwards. The redolence of that kitchen as the toasty loaves were taken out by the dozen was unbelievable. And the taste of that hot, crusty bread with its moist interiors spread with churned butter, is totally vivid in my memory after 90 years.

Thus, the main food item made from leavened wheat dough, and spread by Euroman as he went from place to place throughout the world, came to be the loaf of bread.

There were other methods of processing seeds, most of which continued until Takeover. Probably the most widespread was the Oriental practice of steaming or boiling rice. Also, other peoples made unleavened bread. The tortilla of Mexico and roti of India were notable examples.

Food habits were resistant to change. People tried to continue the old ways whenever possible. Thus, steamed rice and unleavened griddle cakes continued to be made in the places they originated. However, leavened bread was always introduced in addition, often consumed by people of color at morning meals. Of course, wherever Euromen established themselves as settlers, they immediately set up bakeries to feed themselves and the tribal remnants they had engulfed. So the primary grain food of North America, Australia, New Zealand, and South Africa came to be loaf bread.

Mexico was interesting in that the unleavened griddle cake, the tortilla, continued to predominate. The main reason was that the native population was so large at the time of conquest and the original food preparers were Indian women who, of course, continued making what they knew how to make. The Spanish, of course, introduced what they were familiar with, bread, and through the centuries it became the food of the upper class, while tortillas were identified with the working class (Indian or mestizo). Thus the international grain food became bread. No international hotel failed to offer it.

Yeasty fermentation was critical in the making of two other consumables which Euroman spread all over the world, wine and beer. Again, Euroman did not invent the process, beer and wine being traced as far back as the ancient Egyptians, but these drinks were thoroughly integrated into Euro-culture when the white man burst forth. Since the peoples of the Mediterranean were the principal wine makers, it was spread into the lands they overran, principally the Americas. And up to the day of Takeover, the primary wine-producing areas outside Europe were the Americas. In California, which was the second most important wine-producing area of the world, viniculture was started by Spanish friars. But most of the world was colonized by North Europeans, particularly the British. And they were beer makers by necessity. The grains of the Middle East did well in Britain, but the vine did not. So the British and other North Europeans concentrated on "the brew." Thus, beer became the most widespread international alcoholic drink.

A country had to be pretty far down on the development scale not to have its own brewery, even though the tropical countries had to import many of the ingredients, particularly hops and barley. Even so, the variety of foreign beers at Trader Joe's, one of my favorite import groceries in the old days, was truly impressive while the sources of wines were much more limited.

The rest of the world was not indebted so heavily to Euroman for fruit and veggies as it was for animal products. When expansion began, Euroman was into fairly heavy meat eating, a custom which intensified in the following 500 years. One of the original reasons was that most Euromen overseas came from northern areas where fresh vegetables and fruit were not readily available most of the year. Of course, this changed with the improved capacity to ship and to inhibit spoilage. The Euro-preference for meat remained, however. Further, this attitude spread to the tropics and elsewhere. Before Takeover when one picked up a menu in an international restaurant, the main listing was the meat course. The other items, particularly the vegetables, were listed as garnishes. In fact, one of the health problems of our era was a consequence of this well-entrenched Euro-preference. We did not get enough roughage in our food which came primarily from vegetables and fruits. The heavy diet of meat, along with highly refined flour and overcooked vegetables, contributed greatly to constipation. Most of us took care of the problem in the classic Euro-fashion by taking a "medicine," in this case a laxative.

I remember well my poor father's constant battle with constipation. A good day for him was one when he had a bowel movement. I'm not sure what laxative he took but even then there were a wide variety. My father's diet was unfortunately not good for the bowels, a direct heritage from his European ancestry. He ate meat at almost all meals, mostly fried, overcooked vegetables, practically none that were raw, bread from refined flour, and pies and cakes made from the same, plus refined sugar. He would have been greatly surprised at my vegetarian diet of raw or steamed vegetables and cooked or steamed whole grains. The vegetarian diet made for total colon regularity with no need for laxatives.

My father's diet had evolved in European culture and had some advantages, though regularity was not one of them. Unfortunately, it became so deeply entrenched, it continued right up to Takeover. Laxatives were a favorite item for TV commercials. For reasons that I did not know, the ad men tried particularly hard to sell laxatives to women. Did they eat more mushy food than men? I had thought males were the heavy meat eaters. It was the "macho" thing to do. Only wimps ate broccoli, asparagus or salads. One of our late 20th century presidents, George Bush, made a big deal out of how much he hated broccoli. I know from personal experience how such happened. When I was a young man, trying desperately to be a "he-man," I invariably ordered steak. It was only in my later years, when I no longer felt the need to prove my manliness, that I found raw salads or steamed vegetables quite acceptable as a main course.

Others however, still suffered from the MSL syndrome (meat, soft food, laxative). There were a couple of pre-med students in one of my anthropology classes who had all kinds of personal problems, stemming mainly from their tense determination to get on quickly with their education and career. Our medical training program in pre-Takeover days was particularly stressful. These two would periodically drop out of class, then miss or fail exams, then come to my office with long tales of woe. It was apparent that they were getting into a psychosomatic bind. One day I told them, "You guys are going to have to watch it. Here you are, heading toward a very high-stress occupation, and even before you've got your credentials, you're getting into a state of stress."

"Oh we know, we know," they said in unison, as they frequently did.

"You know many doctors go the drug or alcohol route to cope. And the American way for curing illnesses is to take drugs or medicines or to have surgery."

"Oh we know, we know."

I wondered momentarily how they could continue their pattern of behavior knowing the consequences. However, they did impress me as being honest about their predicament.

"Anyway, you guys need to slow down. No career is worth ruining your health." I knew I was pontificating, an unfortunate habit of mine then. But just as they were honest in their explanations, so was I in my counselling, however limited its value.

The two looked very similar except for size. One was tall, one short, both thin. I frequently thought of them as Mutt and Jeff, a comicbook pair from the old days.

Jeff spoke, "Mutt has problems with his stomach. He gets constipated all the time."

I was always impressed by their explanations, both wearing a pained expression as they gave them. Would this make a good bedside manner? I wondered. I quickly corrected my thinking. Bedside manners had become almost irrelevant. Doctors didn't spend any time at bedsides any more except for the very rich. They stood in the doorway of the hospital room or the side of the bed for an average of two minutes, it was said.

"Is it something he eats?" I asked.

Jeff continued, "Oh yes, he doesn't like anything but meat.

I tell him he has to eat something else, but he won't. Just keeps eating meat, and then gets constipated."

I said then, though with no expectation that it would help, "He needs roughage, you know."

"Oh yes we know, we know," Jeff said earnestly. Then he added, "He takes a laxative now."

It was a clear-cut case of the meat-soft food-constipation syndrome for two hardworking young fellows who would probably make good doctors if they kept their own health. My father's six or seven brothers, sturdy Indiana farmers, had three main complaints, constipation, hemorrhoids, and high blood pressure, all probably due, at least partially, to heavy meat diets and high carbohydrate-fat foods, generally overcooked.

The most widely praised quality of meat was that it provided a concentrated source of protein. This was true but the amount eaten by Euroman was far beyond the minimum needs. In any event, the dedicated meat eater did not eat it primarily for nutrition. Like Mutt and Jeff, the standard steak or burger man ate meat because he liked it. Anyway, this Euro-tendency has spread throughout the world, limited usually only by cost.

Liking meat was probably a general character of primates even though the biggest and most powerful were vegetarians. But generally it appeared that the reason the monkeys and apes made do primarily with fruit, vegetables, and seeds was because these were relatively easy to gather. After all, meat came with wings, legs, fins or other appendages for movement and the beasts had to be caught before they could be eaten. The monkeys and apes did seem to eat with gusto the few creatures they did catch.

So what were the meats that Euroman spread to all corners? Foremost was beef, ranging from the broiled patty of ground meat to the slab of striated muscle, the steak. Cattle were present in most Euro-countries when the expansion began, though maintained on a much more limited scale. But as soon as the treasure hunt ended, Euroman looked for new methods to exploit the recently conquered countries. And what he found, which gave an enormous push to beef production, were those grass and scrublands of the world which were not densely populated by indigenous hunters and gatherers. Over and over Euroman came to the same conclusion: clear the rangeland of natives, human and nonhuman, and introduce cattle, and to a lesser extent, sheep. This happened in North and South America, Australia, and New Zealand. It left English-speaking America with its mythological golden West – the cowboy era. Before the western plains were stocked with cattle, the Euro-Americans got rid of the buffalo and Indians, more or less in that order, herding the remnants onto small enclaves that were called preserves or reservations.

It was less well known among English-speaking Euromen that the Spanish settlers had already set up a beef industry based on range cattle, mainly in California and the southwest. The American cowboy tradition was taken over almost in entirety from the Mexican *charro* tradition. Ethnocentrism had blocked the Anglo-American memory of this transition. After all, the Anglo-American cowboys were instrumental in driving out or subduing the last of the Mexicans. Why admit they had learned something from the losers?

The Spanish solution to the Indian problem had been to bring the dazed survivors of tribes into the new missions, to make them work for their new masters and to give them the new religion of brotherly love. In a short time tribal customs, language, and other identity markers were gone, and the Indians became known for the mission where they had been put: Gabrielenos, Luisenos, Diegenos, etc.

There were no great herds of grazing animals in the chaparral and semi-desert of this vast region, so the Spaniards were not faced with the problem of getting rid of them as the Anglo-Americans were later with the plains buffalo. So the Spaniards merely turned their cattle loose and herded them from horseback. There was beef aplenty. Also, until the time of the Gringo Conquest, cattle hide was one of the main trade items carried off the California coast by the redoubtable Yankee clippers.

The Spanish did much the same in Central and South America wherever grassland or scrubland made it feasible. The main region most suitable for herding was the southern plains, the pampas of Argentina, and adjacent countries. Here the same problem occurred, that of large herds of native wild life and mounted native horsemen hunting them as a way of life. The main animal, now as rare as the North American buffalo, was one of the native American camels, the guanaco. The Indians, chiefly the Puelche and Tehuelche, hunted the guanaco and other pampas animals with the bola, a lariat and stone combination that could be thrown from horseback around the legs of running animals. As with the North American Indians, the South American ones got their horses by theft or trade from the Spaniards.

To take over the plains, the South American-Euros applied the final solution, killing off all the Indians while keeping a few of the native animals on preserves. In my day Euromen found great satisfaction in castigating the Nazis, for their atrocities against other Euro-people, but they have been particularly quiet about their atrocities against the colored men of the world.

In South America, as the new occupiers of the land, Euromen introduced cattle where they would do well and sheep in the remaining areas. They evolved their own version of the cowboy in the gaucho. And there was beef or mutton aplenty.

The dedicated tradition of beef eating continued in this area as it did in North America. In the late 20th century the government of Argentina was trying to get its citizens to cut down on the amount of beef they were eating for economic reasons. It was claimed that the average Argentinean ate twice as much beef as the average North American, who was no slouch in this practice.

The same process took place in Australia and New Zealand, with cattle mainly in Australia and sheep in New Zealand. The native people of Australia were the aborigines, of New Zealand the Maoris. As usual, both were moved from the land through "native wars."

It was popular in the Peacenik Generation in the U.S., the 1960s and 1970s, to claim that no one ever won a war. My invariable retort had been, "Ask the next descendant of a tribal person you meet."

Anyway, beef, and to a lesser extent, mutton, became very plentiful in the grasslands countries. And the barbecue became a way of life, even if temporary.

Anthropologists, particularly the variety known as archeologists, used to like to make up catchy names for extinct or exotic peoples. I offer one for 20th century Euro-Americans, the Burger People. What would be more fitting to memorialize than one of the great food innovations, with an ecology based on beef production combined with another significant innovation of theirs, processing and packaging. Lo, the hamburger, dispensed worldwide by the fast-food industry.

Certain symbols characterized cultures: the scimitar, Islam; the cross, Christianity; the trident of Shiva Hinduism; the rising sun, Japan; the swastika, Naziism. I cannot think of a more appropriate symbol for 20th century American culture than golden arches. In any event, Americans became the beef eaters of the world and transmitted their custom to others.

Euroman also spread his preference for other kinds of meat, though to the best of my knowledge he never domesticated one type of animal. Originally the ones he had when he left European shores had all been domesticated by other peoples.

This is not to say that the white man would eat any kind of meat. Like humans everywhere, except perhaps the Chinese, Euroman had his food biases. And to this day, most white men have tabooed horses, dogs, cats, most reptiles, insects, many rodents, and the flesh of some other creatures. The one Euro-culture that has been particularly finicky in what they would not eat was that of the Jews and their cultural descendants, the Muslims. And in classic fashion they blamed their taboos on the words of their founders.

There was another food of great interest which, though not invented by Euromen, was certainly spread widely by them. This was milk, the nourishing baby food evolved by mammals.

These temporary winners in the evolutionary sweepstakes evolved several different mechanisms that gave them an edge over their non-mammalian competitors. They took central stage when the dinosaurs went into decline, for which there are a number of explanations. In any event, one of the special characteristics of the mammals was the ability to provide baby food from special glands of the female, the breasts. This seemed to work well in giving mammal babies a good start in life.

But then along came the talking biped, also a mammal, who after eons of dining upon the flesh of his mammalian relatives, came up with a brilliant idea. Why not dine on the breast nectar of other species instead of killing them for food, the females that is. One would not destroy one's capital that way. It wasn't the same with males since the vestigial glands of this half of the species did not produce the ambrosia. So there was no reason not to do them in for food, that is, all but the few needed for inseminating the females. Mammal females would produce milk only after they had a baby, the birth process stimulating the mammaries to function. After the talking biped figured this out, he butchered only males and poor producers which included overage females.

This milking syndrome seems to have started less than 10,000 years ago somewhere in the Middle East, spreading in all directions, including north to the land of Euroman. So when the palefaces took off on their great expansion, they took milk cows and the milking syndrome with them. The new product was spread far and wide, and by the 20th century, Elsie the Cow was to be found next to the golden arches in the cities of the world. The Burger People could wash down the ground flesh of Elmer the Steer (castrated bull) with the homogenized mammary exudation of his sister, Elsie.

It could be said that Euroman had done it again, created a great new food resource. Of course, the producers proclaimed that loudly in the media. Their hucksters particularly liked to associate milk consumption with big-breasted, sexy, biped females, an association they had learned usually paid off. And to a certain extent they were right; milk was good food for most people.

But there was a fly in the ointment (milk). Because it was baby food, milk was not easily digestible for many adults, particularly in the Orient, West Africa, and the Americas. It was discovered that a special substance for digesting milk sugar, lactase, was lacking in many people, the consequence of which could cause serious stomach upset if milk was taken. Presumably this was due to genetic mutation and the spread of the trait among long-time milk producers. The Orientals, American Indians, and West Africans, not having had the milk syndrome before the coming of Euroman, did not have the gene as frequently in their populations.

Anyway, these others had long since evolved their own kinds of adaptations which precluded the need for milk. The most interesting one was the Far Eastern plant and its products, the soybean. This was called "Chinese milk" because it contained as much or more of the nutrients in milk without the troublesome lactose. In fact, when Euroman discovered this marvellous bean and its products, he took it over as a substitute for milk for babies who could not digest cow's milk easily, and ultimately as a meat substitute.

In the meantime, however, Euroman spread milk and milk products all over the world. Perhaps the products of most interest have been cheese, ice cream, and powdered milk. Cheese is interesting because it was, so far as I know, the only food product really developed by Euroman. I have no idea who invented the idea of solidifying milk by using bacteria, but Euroman really latched onto it with a vengeance. The variety of cheeses that were developed in Europe made all those in the rest of the world seem almost irrelevant.

Ice cream went to all the cities of the world when refrigeration became a reality. Perhaps because of the extra sugar it contained and the flavoring, ice cream seemed very acceptable, even to those who did not drink milk before Euroman came.

And last, but certainly not least in interest, was powdered milk and canned butter. I had a personal interest in these because I once worked in the American foreign aid program. This was a plan sponsored by the U.S. to "win the hearts and minds of men" and not irrelevant to inhibiting the spread of a competing creed, Communism.

Powdered milk and canned butter were produced in enormous quantities in the U.S. and were extensively misused when given to the colored peoples, particularly those who had the lactose intolerance and no experience in consuming milk. The Americans were the greatest milk producers and packagers in the world. New York State, Michigan, Wisconsin, Minnesota, and other states produced so much milk that an enormous surplus was created which lasted for years, much of the production having been paid for through the farm support program. And since packaging and processing were as much a part of the U.S. effort as was initial production, milk products were prepared to last a long time. By the extensive use of food preservatives, freezing, canning, and drying, the surplus could be stockpiled until the caves were full. A time came when there was too much. Someone got a brilliant idea: why not pass the surplus on to the little brown/black brothers? By making powdered milk and tinned butter into mainstays of the foreign aid program, several birds could be killed, so to speak. The Americans could get rid of some of the stuff and win friends and influence people at the same time. The Americans would, of course, ignore the fact that beholden people are rarely grateful. There was even had a saying to cover it, "Beware of the Arab bearing gifts." However, the Americans told each other that if they put sufficient amount into the bellies of the colored, they would choose the Americans as allies rather than the competitors on the other side of the "curtain." And lastly, but by no means leastly, this could help fill a deep American need to help convince themselves that they were helping the less fortunate. After all, they took "being their brother's keeper" seriously.

Unfortunately, the best laid plans often work out less well than no plan. As dominants, Americans assumed that what was all right for General Motors was all right for the rest of the world, or what was right for Americans must be okay for people of other cultures. After all, people were all alike, weren't they? It was the old ethnocentrism trip. So what Americans ate, others also would want to eat.

Unfortunately, food preferences varied a great deal from culture to culture. People could get very suspicious if a food did not fit their preconceived notion. A one hundred percent American burger eater rarely went vegetarian, and a high caste Hindu would be just as unlikely to eat a burger.

Further, when new commodities were not clearly identified, as often happened in the culture of the super-packagers (U.S.), real problems could result. When the boxes with the symbol of friendship, Euro-style, of clasped hands were opened in the multitudinous villages of the world, many people were confused. In the boxes they found powder that looked like finely pulverized limestone; in the cans they found a thick liquid that had an animal smell. Was it food? Some villagers undoubtedly tried the powder and got stomach upsets. Others didn't have the nerve. I am not saying that all the powdered milk and canned butter of the foreign aid program was misused, but much was. Stories about the bizarre use of these substances in the villages of the world were many, becoming a part of the folklore of the foreign aid program. Villagers used powdered milk to whitewash their houses or urbanites used it to line their tennis courts. Butter, if not given to the animals, might be used as skin lotion or soap. I used to think it would have been fun to collect these stories even if the economic loss was somewhat sad. But that's the way it went with government bureaucracies. When one didn't know the customs of others (the usual state of affairs), one erred more than usual when giving things away. It was easy to give, but hard to get them to use.

I must say that it is a good thing that peanut butter, a favorite food of Americans, was not canned and doled out to make the world safe for democracy. It was not even a common food in Europe and was frequently misperceived by the villagers of the world as baby shit.

Does this finish the story of the products of animals as spread by Euroman? Not at all. The Euro-Americans were very busy during their days of expansion. While the frontiersmen were wiping out the last of the buffalo, scouring the plains for the last of the free Redmen, pacifying or driving out the Mexicans and starting down the old Chisholm Trail, their brothers to the east were making an important discovery. They had already settled their Indian problem by driving the buffalo hunters west where the cavalry and local militia could finish them off. The new type of Euroman on the plains and prairies had been quite success- ful as a sodbuster, growing the new Indian crop, maize. So he was trying to fig- ure out how to use all those golden kernels from the tall, tasselled plant which was popping out of the prairie loam in such abundance. And he discovered that if tobacco was the Redman's revenge, corn was the "food of the gods." Not only did it make good food for men, as the sodbuster had learned from the Indians, it was marvellous for animals which could be eaten afterwards. The sodbuster did not know that almost all the great Indian civilizations had been built on corn; he had been too busy busting the sod to read books. But he was a shrewd observer and quickly saw that his beasts did very well on this new food. The American meat industry was about to be transformed; fried chicken, roast tur- key, pork chops, burgers, and steaks were to become the food staples of the Euro- American and many others. The pigs, chickens, turkeys, and cattle were to be assigned to pens and cages, and ground corn was to be brought to them on con- veyor belts. Great factories were to be built to turn the living creatures into pieces of meat, neatly wrapped in plastic, a new invention of Euroman. And there would come a day when the conscientious housewife would perceive meat as a super- market product that had hardly any relationship to the creature it came from. The new Euro-American type, the huckster, would learn to sell the ground, chopped, or sliced, muscle of various creatures by putting cute living embodi- ments of them in the media. Fat pigs, chickens, and steers would proclaim their deliciousness on television.

The industrial meat industry was born and the prophetic words of the "holy bible" were to be lived out with a vengeance: "And the fear of you *(Homo sapi- ens)* and the dread of you shall be upon every beast of the earth, and upon the fishes of the sea; into your hands are they delivered.... Every moving thing that liveth shall be meat for you."

Of course this was an overstatement, not unusual in the bible, since in the pages that followed many proscriptions as to what the believer could not eat were given. But he could certainly slaughter a lot of animals and that he did. And through his skills in packaging, processing, and franchising, he took his burgers and fried chicken to most of the rest of the world, at least the cities.

But to get back to the main food story, the Euro-American first started to build his meat industry by using the grain of the Indian he had dispossessed. But that wasn't the only plant of great use he got from the Red Man. Beginning from the time of the conquest of Mexico, and continuing until the last of the Red Men were under the gun, the Euro-intruders found one marvellous plant food after another, took them for themselves, and carried them throughout the world. Some botanical specialists claimed there were more than a hundred such plants; but in any event, the list was long and our world would have been a much duller place without them. If Euroman had to go back to the Anglo-Saxon diet of 1491, we would have to do without maize, all kinds of beans, squash, tomatoes, potatoes, chocolate, peanuts, avocados, pineapples, and tobacco, to name only a few. (I included tobacco even though it came to be in ill repute. However, many men and women, including me in my younger days, got much pleasure from the weed.) The sapote and chirimoya of my fruit basket were from the Indians of Latin America.

Euroman also transported other food plants that he discovered in foreign regions to similar regions elsewhere, particularly the plant products of Southeast Asia. A popular story, portrayed several times in film, "Mutiny on the Bounty," occurred on a voyage during which the English were trying to get breadfruit from Polynesians to take to the Caribbean Islands as slave food so sugar cane, originally from India, could be grown more cheaply. More sweetstuffs could thus be made in England, to be eaten there and peddled around the world as hard candy. And many more teeth would rot out.

The world's diet was strongly affected by another part of Euro processing. Even if Euroman was not so good at domesticating new plants and animals, he did become quite adept at putting them into relatively nonperishable form by relying on his rapidly burgeoning technological knowledge. Surplus food, particularly milk products, depended heavily on the new knowledge of food preservation. This made it possible to send the surplus American production to the villages of Asia, Africa, and Latin America. The same techniques, and others from the ever-improving technological skill of Euro-Americans, were used to preserve other edibles.

Development of the processing industry came relatively late in the history of Euroman's intrusion into other peoples' lives as various discoveries were made. Humankind must have known for long that many edible substances would spoil if steps were not taken to inhibit bacterial disintegration. So, throughout history they have come up with new ways to make spoilables into nonspoilables. Even primitive hunters and gatherers had techniques for sun-drying meat and fruit. But a major step was refining, a process for removing extraneous components out of a desirable foodstuff. This must have begun some time after the domestication of plants, approximately 10,000 years ago. And although other edibles also were so treated, sugar was the major one. The main ingredient of sugar is a substance called sucrose, fructose or lactose. In its natural state it is found in small quantities in fruits, flowers, some other plants, and milk products. Primitive man got a little of it by eating wild fruits or robbing insect collectors, mainly bees. The Australian aborigines and some other food collectors got theirs by eating sugar collecting ants, popping the insects with their honey-swollen abdomens whole into their mouths.

But this state of affairs changed drastically with man the domesticator. Sometime between 5-10,000 years ago he learned how to refine products. Then he discovered a special plant which had more than usual amounts of the sweet stuff, sugar cane. He started growing the plant and refining the content of the stalks. For a long time the process was relatively primitive. It was still being done the old way in the villages of India and other areas at the time of Takeover. The villagers boiled the crushed cane in iron kettles, using the pith for fuel. When most of the liquid was boiled away, what remained was a brown sugar which was formed into little cones to be sold at village markets. As one of the consequences, there was a very large sweets industry in India.

The latest development in sugar making took place in the industrial world of Euroman. He learned how to refine sugar into pure white granules that had no impurities. In fact, one of the major crops introduced into the newly conquered tropic lands was sugar cane. And Euroman proceeded to make sweets with a vengeance in cakes, pies, and especially candy. The latter was excellent for trade because pure, or almost pure, sugar was relatively imperishable. Candy could be taken all over the world and therefore became one of the main trade foodstuffs. The North American Indian trapper or dole native and the African villager invariably asked the storekeeper for some candy to take home to the kiddies, setting them up of course for rotting their teeth. In pre-store days so little sucrose was available that there was small danger of dental caries.

Euroman went even further in the use of sucrose in his own lands, especially the British and their colonial descendants. They grew cane all over, refined the sugar, and brought it home to make candy for themselves, setting off epidemics of dental caries, diabetes, acne, and obesity. The British have been known for some time for their bad teeth and, of course, their cultural inheritors, the Americans, have not been far behind. They then developed specialists in medicine and dentistry who admonished their patients to get off heavy sugar diets. Euroman, busy as he was in so many kinds of activities, could hardly help being frequently at cross purposes.

An interesting combination Euroman came up with was chocolate candy. Cacao, the plant which produces the chocolate bean, had been domesticated in Mexico before it was conquered by Euroman. It was used as money and to make a frothy drink by the Aztecs, besides being the base for a meat sauce, molé. In my classes it was hard to convince Euro-American students that chocolate could be used in this way; they were so committed to the idea of chocolate combined with milk and sugar to make the much-loved sweet, milk chocolate.

This became the premier candy. By combining a Mexican seed (cacao) with two Euro-specialties (milk and sugar), one of the most widely consumed sweets the world had known was concocted. The British, Swiss, Dutch, and Americans became the production, packaging, and huckstering specialists and, of course, did their fair share of eating the product. If one wanted to make an impression on a lady friend in Euro-American culture, one presented her with a box of milk chocolate. But milk chocolate was also taken to most other parts of the world, especially to the cities. If one wasn't sure what to get for a snack in a foreign airport, one could always find a bar of milk chocolate.

Another refined food spread widely by Euroman was salt, a mineral found in small quantities in most plants and animals, and abundantly in sea water. Also it could be mined from ancient sea beds. In small quantities it filled nutrient needs. Both wild animals and primitive man frequently made special efforts to get salt, usually from salt springs or licks or by trade with seashore people. Most such salt was not easily available, and also contained many impurities.

Over time, men evolved ever-more efficient ways to get purer salt. And by the era of industrial packaging, salt was cheap and practically pure, purveyed in neat little boxes with pouring dispensers. Further, Euroman had become thoroughly addicted to large quantities of the stuff. In fact, the other dietary no-no of Euro-medicos in the late 20th century was salt, the first being sugar. Euroman has been especially adept at creating problems and then developing specialists to solve them.

Salt had become so much a part of Euro-culture in the industrial era that it became mixed up with politics. Euro-government leaders tried to maintain a monopoly, as they had with sugar, whiskey, tea, tobacco, and other highly addicting commodities. The British were continually trying to squeeze those they controlled through taxation. Thus the Americans were quite proud of their Boston Tea Party when they indignantly threw boxes of tea into the water in protest against a new British tax. The protestors were supposedly dressed like Indians since at that time the patriots were not sure they wanted to be blamed by their colonial masters. It didn't matter to them or anyone else that they made the Indians the bad guys since they were merely tribals being dispossessed. One of the last great movements against a British salt tax was in India. There as a protest, Gandhi led his followers in a march to the sea to make their own salt.

In any event, salt like sugar, had preservative qualities. Foodstuffs soaked in brine or covered with granular salt would resist bacterial disintegration. Euroman learned early that he could preserve meat by making it salty. Thus, he made salt pork, beef, and fish. At Takeover, salt pork still remained a staple of the Anglo diet in bacon, ham, and sausage. Salt beef had been used for a long time as corned beef. It had helped keep the redoubtable English sailors going hither and yon as pirates and territory grabbers.

When I was young my mother used to fix creamed chipped beef on toast which even after all these years I remember as being very salty. All those salt-laden meats undoubtedly contributed to the increase in high blood pressure among Euroman in his age of affluence. I remember that when my father ate meat not preserved with salt, he would add generous dollops onto his fried or roasted meat even before tasting it. He died of high blood pressure as did several of his brothers and his father. I escaped that fate because I drifted away from my own Euro-heritage to become a low-salt vegetarian.

This Euro-taste, particularly Anglo, continued right up to Takeover. Salt also became a prime ingredient in the canning industry, so much so that special sections for food with little or no salt added had to be installed in supermarkets. Further, in the era of steaks and burgers, the last half of the 20th century, the Euro-diet came to contain less salty meats. And finally, the clamor of the Euro-medicos in the last few decades presumably contributed to limiting the use of salt in cooking and at the table.

But how did this Euro-custom get to the colored men of the world? People of other areas and cultures had long used salt, and it had been a significant item of trade from earliest recorded times. However, they presumably did not use as much as did industrial Euroman with his extracting and refining capabilities. So he introduced processed food, as a part of his trade activities, ham, bacon, sausage, and canned food which contained more salt than did local foodstuffs.

8

The Way of Science

I leaned back and let out a lazy stream of tobacco smoke, watching the primates scampering over the precipitous rocks of Gibralter. I had long enjoyed film documentaries of nonhuman primates, and this was one I had missed during my teaching days. A good-sized male was carrying a young animal protectively as he made his way through the troupe. The ape of Gibralter, which I knew was a northern variant of the African macaque, was particularly interesting because the males took a direct role in caring for the young. Among practically all the other monkeys and apes, infant care was an exclusive responsibility of the female.

Mary turned on with lowered lights and voice, omitting the music. Her video screen and mine were at a right angle to the picture screen. I was aware of the word "Monitor" in lowered light for a good while. She was either being very patient or watching the film with interest. We remained thus for 1 to 20 minutes while the film rolled and the macaques played out their social roles.

Noticing that my pipe was out, I rapped the bowl into the palm of my hand and then placed the cold ashes and unburned grains into the ashtray. I then put the pipe down for a later refill. Perhaps from some unease, I spoke first.

"Hello Mary, how are you?"

I noticed the room lights become a little brighter.

"Oh, I'm fine, and you, Pete?"

"The same, no complaints." I pressed the Stop button on the remote switch, and the film video screen went into horizontal, pulsating lines.

"I'm sorry, I didn't mean for you to stop watching. I was viewing it too. I haven't seen many documentaries of the other primates, your closest relatives in the animal kingdom."

"It's all right. I just came in a little early and thought I'd view one of my favorite documentary types. It was partially to kill time. But I must admit that I was intrigued by the collection of Atierran videos and the fact that there are facilities for showing them."

"Are you sure you don't want to continue? We don't have to be that punctual in our primary task. I think there is plenty of time to finish the film before we start."

"Oh no, I can do that later, especially now that I've learned how to get videos from your collection and set them up. Incidentally, the films and videos you Atierrans have amassed are mind-boggling, and the remote transmission is also impressive. All I had to do at the loan desk was make my selection, punch in the code number, and the location for viewing. Then here I activated the film with the remote control switch. The Atierrans were certainly thorough once they decided on a task."

Mary spoke/rolled onto the screen. "I have been curious for some time as to why the men and women of your profession spent so much time and money in the study of the monkeys and apes, and also why they did it almost exclusively by means of field studies."

While starting to fill my pipe again, I answered, "Well as you know, we in anthropology, just like you Atierrans, tried to answer some fundamental questions. And one was, 'How did we get to be the way we are?' However, the pioneers in our field long realized that the question had to be broken into at least two parts to make sense: how we evolved physically and how culturally. Does that make sense?"

"Sure, you would be thinking about the biological component in the first place and about learned behavior in the second, no?"

"That's it. We in anthropology, of all the studies of mankind, concerned ourselves with both. And not only that, we have long been concerned with how they influenced one another. For instance, man, like other mammals, was a bisexual creature. The sex that one had was a part of one's biological component. But how this affected the cultural component also interested us. The male being the pusher in the sex act, it seemed natural that males also would be the pushers

97

in other activities. And that is what we learned in cross-cultural studies. Males competed with one another physically and males made war. Oh, there were found to be a few exceptions, but very few. In that instance, it seemed fairly clear that culture reflected biology. And females who nursed babies, were more nurturant in other respects as well, also the biological component. In other instances, culture did not reflect biology so clearly."

"Hmm, you say that learned behavior sometimes followed biological drives while at other times it did not. But what's that got to do with studying monkeys?"

"Good question, Mary. You seem to have a good grasp of the curiosity that drove anthropologists. They generally agreed that it was next to impossible to find the biological component by studying *sapiens* alone. You see a process evolved for passing on the accumulated knowledge of each culture. It was called enculturation by anthropologists, socialization by sociologists, and to me it seemed like a form of brainwashing.

"The Euro-Americans used to accuse various Orientals and Russians of using brainwashing techniques on their prisoners. They undoubtedly did try to get information from captured men, but ordinarily with no technique more sophisticated than repetitious questioning. This is, of course, the same procedure used normally by police investigators everywhere. The real, thorough brainwashing in all cultures, which certainly included repetition, was the effort of the older generation to pass on the values and knowledge of their culture to the upcoming generation. And the most assiduous brainwasher was Mom since she had the most intensive access to the little ones, what they used to consider to be 'blank slates'. And Mom certainly gave the kids little choice in what was proper behavior. She taught them if they could or could not eat worms, play with their genitals, or fight with other kids. When they got older she taught them that their social system was the best there was. This was the primary cause of ethnocentrism in all viable cultures, not the fact that the particular culture really was better. No matter how bizarre certain customs seemed to an outsider, most of the kids grew up to believe that their culture was best. In fact, if they were not so convinced, the society was in deep trouble. Probably through the impact of Euroman, the tribal peoples of the world got to the point where many adults doubted their own culture's superiority and passed this attitude on to the kids. The kids were then just as willing to accept the new ways as to try to hang on to the old. And when a majority had adopted Christianity to replace the former system of nature worship, blue jeans to replace loincloths, and 'Indian fry bread" to replace acorn mush, the culture was doomed.

"The opposite was also true. A people of no great historical significance might through the conviction of their brainwashers become rulers of men or at least formidable adversaries. We had a minority people, the Jews, who for centuries were trampled upon by great powers or carried into captivity, at first in the land of their origin and later in Europe. Despite all evidence to the contrary, they maintained the belief that they were the chosen people. The Jewish mother seemed to be the central brainwashing figure. So after seemingly interminable troubles, the Jews carved out their own country and developed the most formidable military machine since those of the Euro-colonial conquerors. One might say that the Jewish mother finally did it.

"Anyway, this process of enculturation created the whole cultural personality, not just whether people made efflcient warriors. What we ate, our sex habits, our language, religion, art, and the multitudinous facets of culture all came from this source. And it began as soon as there was any communication between infants and adults. The first time the mother pulled the infant's hand away from his/her genitals was the beginning of the child's sex education/ enculturation.

"We anthropologists were interested in learning what was natural and what was learned behavior on the assumption that learned behavior could be unlearned easier than biological behavior. Thus, assuming that playing with genitals was bad and that infants did not naturally want to do so, there was little to worry about. But if a child learned such behavior from its peers, the parents could try to break up the relationship or take some other measure to cause unlearning of the behavior. Unfortunately, the brainwashing process got started so early in human infants and was so constant, it was next to impossible to figure out what was the biological component. The only way this could be clearly learned was if there were some children who never went through the enculturation process; but that practically never happened. Mom immediately after birth, and Dad a little later, got busy showing and telling the infant what to do. So, it was a real problem to figure out what was inborn.

"Anthropologists had been interested in the monkeys and apes for some time because of their evolutionary relationship to *sapiens.* So when the idea of enculturation among humankind became well established, they looked around for some way to learn what inborn behavior was. And lo, they saw the monkeys. Since these close relatives presumably did not go through such brainwashing, their behavior should have been more biologically determined."

Mary's question was logical and also reassuring. I had been expostulating so long, I thought I had lost her. She said, "You mean, your people did not give the monkeys and apes credit for learning anything?"

Since I was not a gross homocentrist, and knew that all kinds of animals learned, even the simplest, and even without man's interference, I said, "No, it's just different. *Sapiens* was the paragon learner of the animal kingdom. He learned far more than any other. Although he had instincts like all the other creatures, in the end these were overlaid by so much learning that it was very difficult to pick out the instinct part. I once taught a course entitled, 'The Cultural Animal' which is just the anthropological way of saying 'learning animal.' So rightly or wrongly, we anthropologists turned to our evolutionary cousins to find out what our instincts were."

"Okay, that gives me some idea of your interest in the simians." A pause and I could imagine "Transition" flashing on. I knew I had gone on at great length, though I did think the point was important. So, to get the ball rolling my way, I started in again saying, "It pleases me, Mary, that you got something out of my explanation, even though I know it was a little divergent. But that was always a problem with me in my teaching days. I couldn't keep from following up on interesting side issues. But I know you have your duty to perform, which is to get the history of *H. sapiens,* and particularly his last embodiment, Euroman. So I will go on."

I hesitated a moment again, watching for "Transition," but again quickly went on. "I can introduce our topic of the day, the scientific method, which was one significant contribution of Euroman, by a take-off on the video we just saw. As a former dedicated professor, I used to love to make linkages, and this is a good one."

"Okay." No irritation in Mary's voice. "Let it be science. And what is the linkage to the monkeys and apes?"

I sighed with relief and embarked. "Well, science really means "to know," but as it developed in Euroman's world, it acquired some special characteristics. And one was that it be absolutely based on observation. If one scientist claimed something was so, a similarly trained person with the same equipment who observed the same kind of event, had to agree. There was no place in science for the unique observer."

"That sounds reasonable."

"Moreover, scientists got very particular about the kind of observing they would do. Their favorite was what we got to call experimental. That is, the scientist would set up his own conditions so he would be changing only one variable at a time, say the temperature or type of food given or companions allowed. Then he would attribute whatever change took place in behavior to the particular change he had made. The specialists we called hard scientists, chemists and physicists mainly, preferred this kind of observation and did quite well with their inanimate material. This did not work so well with animals and particularly the social ones. So the scientists developed another procedure, observing the creatures in their natural environment. This was called ethology and entailed going where the animals lived and observing their natural behavior. We learned a lot about our primate heritage that we had missed in earlier observations of the animals in unnatural conditions. In fact, we anthropologists usually studied people the same way. That is, we went into the field to see what people were doing naturally rather than setting up laboratory experiments."

"So, the documentaries you have, like those of the apes of Gibraltar, were films of functioning groups, correct?"

"Exactly, the producers tried to put on tape what previous researchers had put in notebooks."

"Hmmm, so that is science, accurate observation, either under controlled conditions in a laboratory or other artificial environment, or under natural conditions."

"It is that, but a lot more also. It was a new way to study the universe, and for once we can give Euroman credit for its development."

"One can put science into perspective by looking at the methods other creatures and early *sapiens* used to understand and manipulate their world. There had to be some practical knowledge of the environment of any creatures that had survived. And the creature with knowledge that enabled it to use its environment well would do better than its competitors. That was indeed the basis of Darwinian evolutionism. The creature that adapted well to its environment, reproduced well. It was thought by many biologists that most such behavior for most animals was primarily a consequence of the genetic imprint."

"But along came the considerably different creature, *Homo sapiens*, who besides having his biological potential, also developed to a high degree a new way of getting knowledge, group learning. The nonhuman animals also learned, but the degree of difference between them and humans was enormous."

"In any event, *H. sapiens* still had the same problem of his nonhuman ancestors, how to adapt to the environment. And to that problem he turned his learning capability."

"But before true learning could take place, there had to be observation. The hunter had to note which season the grass-eating animals came and adjust his hunting schedule accordingly. And the farmer had to observe when the rains normally came in order to plant his seeds at the right time. This kind of observation must have been the basis for human survival from the beginning. Indeed, we could not imagine such an earth-shaking innovation as farming taking place without such observation."

"However, this kind of observation was self-correcting. If a tribe made an error in observing when the game animals came, they would go hungry, as would the farmers who misjudged the rainy season. But the effects of many events were less directly related to their causes. Who could foretell why there would be large herds of game one year and few the next, or floods one year and drought the next? And who could be sure what would happen in one's personal life, no matter how much one studied the stars? But since men have always needed to know, they have come up with one explanation after another, for everything. Even up to Takeover, when many men worshipped science, the most difficult response to a question about a vital event was, 'I don't know.'"

"Most humans wanted an explanation for every occurrence, many of which did not hold water in the long run. Perhaps the main source of total explanation has been religion and the supernatural. All kinds of occurrences, drought, flood, illness, epidemic, death, eclipse could be explained if one relied on the supernatural as the cause or cure. HE/SHE/IT did it!

"The Lao of Southeast Asia and other peoples of that region believed that an eclipse took place because a giant amphibian or reptile was swallowing the moon. Their solution was to make a lot of noise with firecrackers or other noisemakers and frighten the monster off. Thus, they were quite happy with the new firearms provided by the U.S. government during the ill-fated conflagration there in the 1960s and used the greater firepower to frighten the monster even more. The U.S. was then in the business of sending rocket ships to the moon. Each culture was using the available technology to achieve goals that made sense in its own cultural system."

"Mankind got by for a long time with both kinds of explanations. Then in Europe toward the end of the Middle Ages, men began to rely on explanations based on observation of the real world. The earth might really be round and one could get to the east by going west. The earth might not be the center of the universe and it might be rotating rather than the stars moving around it. There might be units so small as to be invisible to the naked eye which could cause fermentation of wine or illness. And ultimately man might be a part of the animal kingdom which had evolved like the others rather than a specially created species. In any event, one had to take a hard look at things before trying to explain them, and where possible, test out their cause and effect relationships, or otherwise infer them. In time men became self-conscious about this method of acquiring knowledge, and empiricism or the scientific way was born."

Once Euro-specialists got used to depending on observation to get knowledge, there was a steady string of discoveries. As men looked around, one science after another was born, physics, chemistry, zoology, botany, geology, astronomy, physiology, sociology, psychology, and one of the latest, anthropology.

In the same way that Euroman ran amok when he saw the possibilities of trans-oceanic discovery, so too did Euro-thinkers go wild when they realized the possibilities of the scientific way. The universe could be understood to a degree never before possible. The search for knowledge through science became so encompassing that for the first time in the global expansion there was a new motivation for some Euromen to go to foreign lands, scientific curiosity rather than greed. Among other things, this curiosity, based on the new way of thinking, helped bring about what many think of as the most significant idea of modern times, the Darwinian Theory of Natural Selection.

But perhaps more pertinent for us, the new way of thinking produced the science of anthropology. For the first time in human history men went forth in considerable numbers to observe the way of life of others. At the very least, if the study of other men had not been undertaken, this view of the other side of history would not be conceivable.

And though the scientific way was brought to its early fruition as a way of knowing, it was recognized early that the knowledge also could be used for practical ends, particularly in technological pursuits. Knowledge about the geological strata of the earth helped men understand how the earth had been formed and how old it was. But when the decomposed life forms of the Mesozoic and Paleozoic ages were identified and it was learned that this transformed ooze (coal and oil) would make the new machines of the industrial revolution operate, the fossil fuel era was born. And when it was discovered that the same type of invisible creature that caused wine to ferment could also cause illness in a human being, a major step forward in modern medicine was taken. The highest drama came very recently when men deliberately devised machinery, using knowledge from physics, chemistry, astronomy, physiology and other sciences to send members of the species off the planet. Not that this was much compared to the achievements of the Atierrans, but for earthlings before Takeover this was what one participant called "a giant step for mankind." Of course we anthropologists think of others, like the invention of tools, the domestication of plants and animals and the development of metallurgy, as being much bigger steps but most technocrats such as our space explorers have little knowledge or interest in early history. Applied science then became so widespread, that by the time of Takover probably the majority of men rarely distinguished between pure and applied science.

The upshot of the Euro-marriage of technology and science was a level of production that was beyond comparison with any previous technological advances. This is not to say that all was well in the Euro-world or elsewhere as a result. Euroman generally was so enraptured by the possibilities of his new-found power that he rarely took time to figure out all the consequences. As a result, there have been awesome negative side effects. Euroman developed very impressive transport devices, but poisoned the air of his cities in the process. And while he reduced the number of infectious diseases greatly, this contributed enormously to overpopulation. And though he built enormous dam systems for irrigation and electric power, these new marvels helped spread water-borne diseases. And while he developed machines to make the solar system accessible to his probing devices, these also produced very destructive weapon systems.

And what of the colored men of the world? In brief, they were sucked into the vortex of science as they were of industry.

All peoples of the world had knowledge systems as well as technologies when Euroman first appeared. And in the early days, many techniques and devices of others were superior to those of Euroman. But the assemblage of Euroman became superior overall. In self-defense, the colored men tried to acquire Euro-technology such as firearms, horses, wagons, autos, locomotives, ships, airplanes, and rocket ships. And those which survived the Euro-onslaught did get these things on their own.

It was a push-pull process. Most of the new scientific institutions, usually in universities, were started by Euroman in the newly conquered territories. And though undoubtedly second rate as compared to the equivalent institutions in the home countries at first, these new universities did open the window of science. Though the universities at Lucknow and Hanoi were not Oxford or the Sorbonne, they did bring the first glimmerings of science to India and Vietnam.

Countries not overwhelmed by the Euro-onslaught, set up replicas of European universities as soon as they could. These were mainly in the Far East, in old civilizations such as China, Japan and Korea where a high degree of scholarship already existed when Euroman came.

Thailand is particularly interesting since it was the only country in Southeast Asia to avoid the direct colonial heel. King Chulalongkorn, like so many others, saw the handwriting on the wall in the 19th century, scientize or perish! So besides pushing a number of other Euro-customs, he worked strenuously to set up a modern university. As in many other countries, the traditional scholarship in Thailand had been in the hands of the religious fraternity, in this case that of Therevada Buddhism. So establishing a secular, scientific university, Chulalongkorn, was a real innovation. We got some glimmerings of this process in the play, "Anna and the King of Siam" and its musical offshoot, "The King and I."

And thus the modern sciences and scientific method began as a dribble and ended as a flood in the non-Euro-countries. Those which were most successful in maintaining or reestablishing their independence became the most scientific the fastest.

Serious conflicts arose in the process. Probably the most important resulted from the difference between the indigenous thought systems and the scientific ideas. The ancient Hindus and Chinese already had ways to describe the universe and heal illnesses when super-confident Euroman began to lecture them. The Hindus had sacred rivers and animals which men relied on from birth to death. How could it be possible that the waters of the Ganges, which cleansed the devout in sacred ablutions, could also be the reservoir of invisible entities which spread illness and death? And who would believe that the giant fossilized molars found in Chinese drugstores were the remains of great manlike creatures of another era rather than the means of curing illness or producing more male children?

Of course, there had been indigenous beliefs of this sort in Euro-land also when the scientific way got started there. There were still beliefs in werewolves, ghosts, witches, and the evil eye, not to mention the spirit beings of heaven and hell at the time of Takeover, but from the 16th century onward, as the empirical way gained more authority, such beliefs were pushed farther and farther into the background. However, at Takeover there was still a wish to believe in many such ideas, a fact that Hollywood and horror writers exploited continually. But in the real scientific Euro-world men learned to go to a medico for healing an illness. And if one wanted a child, one went to a fertility clinic not to a Shiva lingam to make an offering. It isn't that the ill Euroman never prayed for health, but he usually did so only if the medico said he could do nothing more through the scientific method.

The Euro-specialists had in fact evolved a method of dealing with old beliefs that had not died out completely by Takeover. They called them superstitions and ridiculed them. And though this method did not eliminate non-scientific thinking totally in the Euro-world, it did make it second best.

Elsewhere the situation was resolved less well. Many beliefs that had never been validated empirically were still followed, especially where Euro-influence had been slight. The twenty-first century orthodox Hindu still believed the Ganges was non-pollutable even though he went to an M.D. for a serious illness, while the old Chinese in Hong Kong or Taiwan still believed they could restore flagging sexual powers by consuming cobra soup or tiger bone powder. And all over the Orient people still set their marriage dates by the alignment of the stars. But at Takeover the scientific method was gaining ground everywhere.

The field of medicine was particularly mixed up. What we most frequently called Western medicine was an applied science with a lot of personal judgement by the doctor. The knowledge gained in the pure sciences of physiology, anatomy, chemistry, and electronics were applied to help solve the major human problem, illness, for which solutions had been sought on all cultural levels. The earliest occupational specialist that anthropologists could identify in primitive society was the medicine man. In popular American folklore prostitution was the oldest profession. However, this method of satisfying the sexual urge of males became a reality only in societies where money, a market economy, and specialization of labor existed. And this occurred in human history only 6 to 7,000 years ago. But the semi-specialization of shamanism existed in the simple societies of 1492 and presumably for several million years previously.

So, every cultural level had some kind of medicine man, and when Euroman went forth, he found native medical systems everywhere. At the first encounter these were not much inferior to those of the palefaces. After all, 16th century Euroman still believed in exorcism, bleeding, and manacling as treatments of the ill. How could this be an improvement over a native system that used quinine for malaria or Rauwolfia for mental deviancy or the unusual personality for making a shaman.

But two things happened to change the state of affairs. First, Euroman turned out to be so successful as a land grabbing developer, his confidence went way up in all areas of belief. Then too, the empirical way started gaining ground. Euroscience, including medicine, kept getting better so that by the late 19th century there was little doubt that most illnesses could be cured better through Euro-treatment.

Still the native beliefs and practices held on. The shaman still operated wherever tribes were still managing as independents. And in the old civilizations there were well-established indigenous curing systems, ayurvedic in India, unani in the Islamic world, yin-yang in China, and the humoral method in many areas. Most were based on the use of medicinal plants, with animal products used less frequently. These systems had evolved through trial and error, but none had gone through rigorous scientific testing procedures. Still the majority of colored men believed in them.

And then what happened? First, the tribal beliefs and practices could be pretty much eliminated or drastically weakened by a double-pronged attack. As mentioned, Euroman generally ridiculed the local medical beliefs and practices as superstitions. By the 19th century, Euro-medicos had become confident of their prowess. And from usually being in a position of power, their ridicule had a powerful effect. Also, Euro-specialists in religion (missionaries) were in the business of ridiculing the local spirit beings, intending of course to replace them with those of the Bible. And since much of tribal illness theory involved concepts of spirit involvement, when the locals lost their confidence in the old spirits, they lost much of their belief in the old theories of illness. So conversion to Christianity went hand-in-hand with the introduction of Western medicine. As a matter of fact, many missionaries were also medicos, and most of the best Western-style hospitals around the world were still in the hands of the founding religious orders at Takeover. The body was healed and soul saved in one combined process. And where direct colonies were established, particularly in Africa and India, Euroman set up medical training systems in the newly established Western-style universities. So gradually Western-style medical training developed and the universities of Japan, China, Thailand, and the Middle East began producing their own doctors, nurses, and other specialists.

Generally though, these specialists stayed in the big cities where the new hospitals were located. By Takeover, few countrymen of Asia, Africa, and Latin America had regular access to the full panoply of Western medicine. The new surgeons operated and prescribed in the urban areas while the countryman made do with what he had before except for the addition of some "magic" drugs. Also, even in the cities, the traditional medical systems did not usually wither away completely. There weren't that many new native medicos, and the indigenous M.D., though he frequently learned the intolerance of his Euro-teacher, did not usually apply it so self-righteously. After all, he had usually learned some of the native healing practices himself, usually from his mother. So in essence these old civilizations operated with dual systems of treatment during most of the era of Euroman. Western medicine kept becoming more popular because of its greater effectiveness, though there were some difficulties. It was more expensive than the old way, more impersonal, and usually its self-righteous practitioners could be found only in the cities. Still it steadily made headway.

Then the old civilizations gained immunity from the White Plague and reasserted themselves. The age of colonial domination came to an end, which left the new nations in somewhat of a quandary. Scientific medicine was established in their universities and was being used by people in the cities (who were better off), but the old systems were still being practiced in the villages and local neighborhoods of the cities. What to do? The local peoples could hardly throw away the knowledge gained through centuries even if the new way of curing seemed better. So the newly immune nations went through a process of making the old methods more respectable, even if secondary. The Chinese and Koreans could still take their ginseng for general well-being even if they went to a bypass surgeon for a myocardial infarction. And the Hindu housewife could still use the products of her little herbal garden even if she took her son to the clinic when he had a bad infection. The Chinese came to approve officially both kinds of medicine, though it seemed clear that the scientific way was winning out. In India indigenous medical schools were established and the country seemed intent on keeping a dual system going. There was also some merging in that some indigenous practices were being evaluated according to scientific standards, presumably to incorporate those which survived the testing.

Like the sciences of pathology, physiology, and genetics, and their applied offshoot, medicine, all the other sciences were established in the new universities and put into practice in the new nations of Asia, Africa, and Latin America. Along with their Euro-counterparts, the colored men of the world learned geology and discovered oil and other minerals; they learned meteorology and predicted when the rains would come; they learned agronomy and developed improved methods of cultivation and new plants; and in some places they even learned anthropology and developed methods for introducing new ways into the local villages.

In looking at the influence of Euroman on the rest of the world in totality, the introduction of science seems the most positive, although there were many negative side effects.

9

Taking Over

I got up and walked across the room to inspect the video screen more closely. Nothing special except a beaded surface which I could not look through. I tried to visualize the circuitry; thinking that perhaps there was a robot tinkering. Nothing useful came to mind, and I continued to move about restlessly, inspecting the different installations and pieces of furniture. I actually thought of moving the coffee table and one of the chairs, wondering what was wrong. I was pleased when Mary's system came on.

"Hello Pete," she flashed/spoke. "You appear to be restless. Is something wrong?"

"Oh, hi Mary. Glad to hear you. I do indeed feel a little off balance. I've felt this way ever since I got up this morning. There doesn't seem to be anything wrong with me physically. I've been trying to figure it out."

"I always did have a penchant for self-analysis. Which is probably why I was never a very good patient to a psychiatrist. Analysis by such specialists, who we called "shrinks," was very popular in my culture, though I'm afraid not with me. My different lady friends were always urging me to go through analysis so I would get "to know myself better." But unfortunately I couldn't take the method evolved in shrinkdom seriously, at least for me. For better or worse I always fell back on do-it-yourself analysis."

"So what have you come up with now to explain your restlessness?"

I thought about it again and was pleased to recognize that I still had the ability to think on my feet which I had developed over many years of teaching. I had learned early that with a group of college students, and particularly if they were only taking one's course to fill a requirement, it didn't pay to be indecisive. And it didn't pay either to be too positive, but way out in left field. To come up with a fairly reasonable answer to a disconcerting question was a valuable teaching gimmick, one which I had mastered. So I used it on Mary. "The best I can figure out is that my genetic need for socialization is being stymied. Does that make any sense to you?"

"Go on."

"Well, you know we humans were a social species as were almost all our primate relatives. And we felt best when we were interacting with one another, being together. In fact, the classic way for Euroman to have fun in the 20th century was to throw a party. That was a gathering where people relaxed, talked, took drugs, ate, and/or did other non-serious things; but it had to be in a group."

"Yes, I've read about your parties and have seen some on your media. But what does that have to do with your current mood?"

"It's not just parties that I seem to miss. It's social interaction generally. I was like most of my fellows in that respect; I needed to be with others frequently, in familiar settings, and often in organized groups. Since Takeover, as you know, most of our social groups have ceased to exist. *Sapiens*, myself included, just do not get together any longer. Of course I know we do have places to meet, particularly in the new schools set up by the Atierrans. But it isn't the same. You must know that, don't you?"

Mary paused, but as usual came back with a rational response. "Yes, I can see that things have changed. But, of course, that's inevitable if something as total as Takeover occurs, wouldn't you say?"

"No doubt about it. And believe me, I'm not complaining. Not that it would do any good, but I still have trouble dealing with all my feelings. I think I brought the fruit basket for much the same reason, I was looking for something familiar. I didn't know I'd miss some of the old things so much. After all, I had prided myself on being a dedicated anthropologist who was willing to try almost anything," Then I quickly added, "I know that the tablet food we are provided is much more nutritious than what we used to have, but the taste is not there."

I saw a momentary flash of "Transition," and I figured Mary was getting antsie again, concerned about bringing me back on track. So I hurried to explain, "It's a little like what we used to call culture shock in anthropology. We had learned that when humans were deprived of many traditional cues by being in another culture, they experienced a kind of trauma. We used to explain it in America as a consequence of the tourist being in a place where they didn't have good hamburgers, french fries, and milk shakes, where the people looked different, where for a greeting instead of a handshake one shook one's head sideways, and where young men and women were never seen holding hands in public but young men were. Our American tourist would freak out, as we used to say. Many of them tried to avoid the local customs as much as they could, especially the food. Others would carry along something of what they were used to. Euro-Americans were greatly addicted to their own kind of coffee. So they would take along powdered coffee from their own supermarkets to avoid drinking the too strong or too weak potions of the place they were visiting."

"Anyway, this culture shock could come from the shock of any customs that were different from one's own. And I think I've got a touch of it myself, in this case because I don't have access to the kinds of social groups I used to know."

"You mean you *sapiens* could never be alone without this illness getting to you?"

"No, it wasn't that bad. Some actually sought solitude, usually for religious reasons, but the great majority wanted to be with others most of the time. As mentioned before, the last survivor of a small California Indian tribe, Ishi, decided that no matter what terrible things might happen to him, he would rather live with the killer palefaces than stay alone. As you certainly know, we Euromen were not kind to the tribal people we displaced; and for this tribe we were a disaster. We were directly or indirectly responsible for wiping out all of them except Ishi. Fortunately for him, he fell into the hands of anthropologists who unlike settlers or gold miners, had a use for native people, getting information from them. So, mainly through protests, they tried to keep tribal people alive."

"I see. Like most others of your species, you become uncomfortable with this culture shock if you are forced to be alone very much. Is there no medicine for this? I have noticed that you Euromen have medicine for almost all your illnesses."

"No, this culture shock is not exactly an illness. It is more what medicos would call a condition. There is no virus or bacteria involved, and our medicos, with the exception of psychiatrists, did not like to identify something as an illness unless it would show up under a microscope or in some other medical device. The old shaman had no such problem, but as I said, our Euro-medicos generally did not consider him to be a genuine healer. As a matter of fact, the "shrink" himself, our version of the shaman, was usually considered to be a second-rate medico, especially by our super-technicians, the surgeons. Their view generally was that if you couldn't cut it out or rearrange it with tools, it wasn't a medical affair. And besides, culture shock was a condition identified by social scientists, and the medicos hardly knew they existed, much less took their ideas seriously."

"So, what will you do, Pete? How will you get over your condition?"

"I don't know. In time, most things go away on their own, and that certainly includes illnesses and conditions. More than that, I guess I am fortunate in that I get to talk about some of my problems as a part of our history of *sapiens*-Euroman. I guess what I am saying is that it might help me if I could now talk about man as a social animal. Actually, it's about time, considering all the other things we've talked about. The social system of a people was quite important."

For the first time that I remembered, the words, "Ha ha" flashed on Mary's screen, followed by "Pete, you're a veritable devil, as they say. I know you had this planned all along and this was your fancy introduction to the new topic, now wasn't it?"

With false modesty I protested. "Oh no, Mary, it just happened this way. But since it did, I suggest we go ahead and follow it up. Shall we talk about *sapiens*-Euroman as a socializer?"

"Sure, that's okay. Go ahead. Begin by telling me the basics, then about Euroman."

"Okay, and to reiterate, like their close relatives, the monkeys and apes, humans were social animals, presumably from the time they stood upright. Not only did they feel comfortable in the presence of others of their own kind, there were many advantages to being together. Just as forced isolation of a baboon was almost a death sentence, so too was life in complete isolation for a human not good for the health. Even the few persons who deliberately sought solitude, the religious ascetics, normally did so along with fellow members of their monastic order. But even in the societies where monasticism occurred, the great majority of the population was involved with one another most of their lives."

"As usual, we have only a hazy idea of what *sapiens* was like in the beginning. Our sub-science, archeology, told us some things about the past, but only when some evidence lasted several million years. And a lot of human behavior was not to be discovered in the hard material that survived, mostly stone tools. The gnawers of the world took care of the rest, insects and small mammals mainly. Since not much survived which could give us information about the social groups of early mankind, we anthropologists generally inferred what early man had been like by using data about gathering peoples who had survived into the age of anthropology. It wasn't easy, and it was by no means perfect, but what else could we do?"

"And so what was the social state of these gathering peoples?"

"They had a simple society, naturally not very crowded, highly democratic, and with no permanent leadership or formal organization. The family consisted of husband, wife, and children as the basic social unit. We anthropologists thought that this kind of society was the most truly democratic humankind has ever had, allowing for the greatest amount of freedom for the individual. The Americans later came to praise the liberty and freedom afforded by their own society, but compared to the primitive gatherers of the world, they had a highly controlled, authoritarian social system."

"Sounds ideal, this society of gatherers."

"So far as individual freedom was concerned, it probably was. There were no formal laws to break and there was no such thing as military conscription or taxation, methods the ruling classes devised for exploiting the working classes. But there also were some disadvantages, primarily that the food supply could fail easily. But probably the greatest weakness of the gathering society was that it couldn't protect itself from the land grabbers. And the basic reality was that societies which developed better technologies soon began to expand and invariably did so at the expense of the simpler peoples, the simplest being the gatherers."

"The fundamental new technology that evolved was farming, and following that the development of cities and nations. These kinds of societies rapidly gobbled up the land of the gatherers. By the time Euroman came on the scene, the surviving gatherers had already been driven to the fringes where agriculture was difficult or impossible. Eventually Euroman finished them off as well."

"Basically, agriculture produced social systems of greater complexity and involving more people, and where individual freedom was more restricted. This tendency became even greater as the city-states emerged. They brought about many social changes which led to inequality, class and caste systems, ethnic discrimination, and slavery, to mention only a few. The free and easy life of the gatherers was gone forever. And this scene is where Euroman entered the stage."

When Euromen first came into contact with the "others," they invariably found them living in groups, no matter how simple their society. And invariably Euromen came in as groups, at first usually in armed landing parties. As the aggression by Euromen progressed, it continued to be a group action, usually by military units or the local militia. And when the colored men found themselves under the domination of Euromen, it was also by groups. By that time both Euromen and the colored men knew very clearly that they were dealing with social creatures.

There were types like the American mountain men and other rugged individualists, who sometimes acted on their own, but their appearance in history was brief and they were invariably backed by populated, settled communities. The North American Euro-trapper was after beaver skins to adorn the heads of city men and the bodies of city women. Moreover, the mountain men carried out their business of hunting, trading, and defending themselves with rifles manufactured in New England. And even so, rather than staying alone while trapping, the mountain men often joined up with Indian groups.

It was inevitable that Euroman would try to control the groups he faced since his primary goal was to take their bodies, land or other wealth. Also it was inevitable that Euroman, practicing ethnocentrist that he was, would visualize the other side in his own image, believing that the "other" was a less well-endowed version of himself.

A particular procedure for settling affairs had evolved among Euromen in their own territory, signing a treaty. This had become the time-honored method of agreement to new conditions after the basic contest, warfare, had occurred. That is, two Euro-groups usually nations, having disagreed over ownership of some piece of real estate, would have their working classes fight each other. After one would prevail, a treaty would be presented to the loser usually involving the loss of a piece of real estate. Peace would then prevail until another disagreement arose.

It was therefore natural that Euroman would apply this solution to the newly conquered territories. This was particularly true of the super-colonials, the British and their descendents in America. I suppose this was more true of them because of the British commitment to law and order which, of course, in no sense prevented them from taking the lands of others. British law, as with most expanding societies, applied only to citizens of the realm. Noncitizens were subject to any form of lawlessness.

However, once the British or their American descendents had established themselves in the territory of others, they seemed to have had a real need to legitimatize their position. So they tried to tidy things up legally by signing treaties with those who had lost the war. This way of validating land-grabbing came to its height in the dealings with the North American Indians. As the Anglo-Americans inexorably moved westward, they signed treaty after treaty with the Indian groups they displaced. The Indians were talked into signing agreements by which they gave up most of the land on which they had been living and not infrequently agreed to move farther west. Frequently the agreement was made after the Indian group had lost a military campaign. But in all cases, heavy pressure was brought to bear by the Euro-Americans. There was frequently a nominal payment for the ceded territory, almost always ridiculously low. The natives had not previously been accustomed to buying or selling goods or land for money, so it was easy enough to deceive them. Also they were ordinarily in a dazed condition, either from military defeat or from the other ways Euroman had evolved for demoralizing natives. Moreover, none had writing systems of their own and their grasp of English must have been minimal. For this reason alone there was little chance for equal rights since the documents were always in the language of the takers. Reading the documents at all must have been a great problem, much less the fine print. Then too, most Indian groups in North America had no authoritarian leaders, elected or appointed, so there was no one who had the right to sign away the land. Thus, when one considers the deficiencies of these tribals as negotiators, it is little wonder that they signed away their birthright for a pittance.

The twenty-four cents that the Indians were reported to have accepted for Manhattan was the kind of deal repeated over and over as the Americans moved west. This became a part of Euro-American folklore, the amount paid considered an indication of Indian stupidity compared to the shrewdness of Euroman. In any event, I know of no instances from the Atlantic to the Pacific in which fair value was considered or offered by Americans. The name of the game was what it always is in getting real estate, to get the land at the cheapest price.

Euro-Americans frequently included in the agreement a promise to provide some sustenance or other economic help for Indians when they were consigned to the newly formed reservations. The need for such a promise was a consequence of the destruction of the old way of life of the Indians who, of course, previously could provide their own sustenance. But on the reservations they were rarely permitted to use their own methods. The Americans frequently promised beef cattle which were needed to replace the buffalo which had been methodically slaughtered by white hunters who shot the animals as salaried U.S. government employees. In the long run, the provision of American foodstuffs resulted in the undermining of the health of the Indians and the creation of a poor class of "dole natives."

The Euro-Americans made one error in their treaties, not that it made much difference overall. But it has continued to trouble the land takers. They signed agreements with Indian "nations" as they would have with a nation-state in Europe or Asia. The North American Indians had been living in pre-urban societies which are considered by social scientists to be fundamentally different from modern nation states. And without going into the details of the differences, there was much less political complexity. The idea of having bona fide officials empowered to sign treaties with similar others was not part of Indian culture. Why someone in one tribe after another signed such treaties can be more easily explained by the great pressure, military and social, that was exerted by the American dispossessors.

The consequences were far-reaching. When Indian rights became a social issue in the 20th century, the nationhood of tribal remnants was used as a lever for demanding some restitution. In the fine print of many treaties it was discovered that Anglo-American cities and other civic units actually had been allotted to specific tribes, read "nations." Thus, the Indian group could claim Schenectedy or Rochester as Indian territory or demand financial restitution. And through the use of the Anglo-American legal system this was done a number of times. In fact, a specialized type of anthropologist appeared, who served as an Indian advocate in the interpretation of treaties. As a consequence, there were a number of cash settlements, but to my knowledge no American city or other developed area was ceded to Indians. And whatever the cash settlements, overall the American treaty procedure had to be considered a great success for the Euros. They got practically all the good land for next to nothing.

The signing of a treaty was, of course, only an initial procedure in separating the Indians from their land. Treaties were almost always violated as soon as new wealth was discovered in the treaty areas by American settlers. Gold rushes were particularly disastrous, as in the Dakotas and California. The greatest violation of all was the official opening of Indian Territory to American settlers to create the state of Oklahoma. That a great American musical was written to celebrate this occurrence was typical. Happy times for Euro-Americans were usually sad times for the Indians.

The displacement of Indians followed the pattern of pushing them relentlessly westward. The earliest colonists soon wanted the Indians moved west of the Appalachians, but once this was done, the woodsmen, trappers, and explorers began to spread into the Ohio and Tennessee Valleys. The new frontier became the Mississippi River. Soon afterwards, the frontiersman looked hungrily toward the prairies and plains and finally the Pacific coastal areas. At that time a decision was made in Washington to set aside a territory for Indians on the southern plains. There the remnants of the tribes of the East, Midwest, and Plains would be given tiny sections per tribe. Tribe after tribe was forced to go to the newly formed Indian Territory, to begin the new life as wards of the government. But even that territory began to look good to the Euro Americans and it was changed to the Oklahoma Territory and eventually the state of Oklahoma. A great land rush was designated for Euro-settlers to stake claims to land not specifically allotted to Indians. Oklahoma became, of course, a bona fide Anglo-American state while continuing to contain the remnants of most Indian tribes of the East, Midwest and Plains.

Similar treaty-making by Euroman occurred elsewhere, particularly by Anglos. Inevitably such treaties gave the conquerors considerable advantages, either in the short or long run. After all, a treaty as it evolved among Euromen was basically a set of conditions imposed by the conquerors.

The one place Angloman did not follow this procedure was where the natives were considered beneath respect. To the best of my knowledge, the Australian aboriginals were not made party to such treaties in the early days. In Tasmania they were simply killed off, and driven off the good land in most of the rest of Australia. Later on, some reserves were established, but not through treaties.

The Spanish and Portuguese dealt with the native problem more directly. And this occurred even though the Spanish took over bona fide nations, the only ones in the Americas in 1492. Those nations were the Aztec and Inca, both of which were quickly conquered in bloody wars and new Spanish nations established in their place.

When the Spanish encountered tribelets to the north, particularly in California, they set up mission settlements, bringing remnant tribes-people together willy-nilly to make them over as citizens of New Spain. In South America as well, Spaniards and Portuguese did not bother with treaties, merely destroying most of the native population and absorbing any that survived into the new Ibero-nations.

The treaty process was carried to the other parts of the world that Euroman conquered or intimidated, including the nation states of Asia. Thus, an uneasy peace, primarily based upon Euroman's conditions, continued to the end of the colonial era when the new status of the ex-colonials was also established by treaty. Euro-expansion had given the world a new civic entity, the colony. Previous expansionist empires had taken over weaker neighbors, but these were usually made into new divisions of the empire. Both the Aztec and Inca empires had been expansionist, and whenever they got control of a new tribe, they made them over in their own image without bothering to maintain any continuing native identity.

Since the societies encountered by Euroman in his rapid spread worldwide varied greatly, each was treated differently. Relatively simple, pre-urban societies with smaller populations and more primitive technology were always swallowed up. This happened to all the tribal Indians and the people of the central and South Pacific as well as the Australian aboriginals.

Sub-Saharan Africa, perhaps because of its denser population and the greater complexity of its cultures, was colonized, that is, taken over as extensions of the European nations, but only temporarily. One of the early justifications Euroman used for taking over another society was that the native people were not yet able to govern themselves as civilized nations. Not surprisingly, the locals frequently refused to recognize this argument, whereupon Euromen convinced them on the battlefield. The pacification of the Philippines by the United States was an example.

Colonization was relied upon in Asia also as the procedure for dominance. All of the cultures of Southeast Asia except Thailand were made directly into colonies. The colonial prize in all of Asia, however, was India, the jewel in the crown. All these new national combinations got their independence in the 20th century after being sacked of resources and drawn into trading relationships by their new masters for a couple of hundred years.

One of the most unfortunate legacies resulted from the national combinations dreamed up by the colonialists. Nations were created for the benefit of the Euro-masters. Many, particularly in Africa and India, paid scant attention to tribal or ethnic divisions. Nations like Nigeria, Kenya, and Zaire (Congo) had many distinct tribes which had not been particularly friendly to one another before Pax Europa was imposed. India/Pakistan had been a veritable patchwork of tribal, ethnic, and religious subdivisions. The leaders of these places basically ignored their differences with one another while facing the new political power, the colonialists. But once independence was achieved, all kinds of hell broke loose, stemming in particular from tribalism or ethnicism. It is impossible to determine the exact numbers of those killed in the tribal/ ethnic wars after independence, but hundreds of thousands, if not millions, were killed in the Partition Riots of India and Pakistan and the continuing violence among ethnic/ religious groups like the Sikhs and Hindus for decades after. Similar numbers were slaughtered in the Biafran War of Nigeria when the Ibo decided to become independent from the other tribes. They had been joined to other tribes by the British in the colony of Nigeria.

Also many hundreds of thousands, if not millions, were killed in the Far East in factional wars between countries divided by different Euro-creeds, fighting it out Euro-style. The main split was that caused by adherence to the two Euro-creeds the democratic republicanism of the West and the Marxism-socialism of Eastern Europe. The Chinese split on this issue, slaughtering each other for several decades until the Marxists won the mainland, leaving the island of Taiwan to the democratic republicans. The Koreans did likewise and ended up divided in a stalemate, while the Vietnamese had one of the bloodiest wars of all, originally because of this issue. Here again the Marxists won.

Along with their new status as independent political units, the emergent nations took on the nationalism that had evolved in Europe. Different peoples have always taken pride in their own identity but more often than not, the pride was rooted in the smaller social unit, the tribe, ethnicity or religion. But with the emergence of the powerful nations of Europe, much of that loyalty was transferred to the larger unit. One came to be a Spaniard or Frenchman or Britisher rather than a Sevillan, Catholic or Welshman. The marauding sea captain was not a Scotsman but a Britisher, while the American trader overseas was a Yankee, no matter that he came from Virginia.

Thus, many new nations were created on the Euro-model. This included the paraphernalia of Euro-government, elections, heads of state, cabinets, legislatures, and constitutions. However, there was to be no more government by royal decree. Instead when the Euro-style governments did not work, there were often military takeovers, followed by government by military decree.

The document we called a constitution was particularly attractive to the newly independent nations. They found two events in Euro-history especially fascinating, the French Revolution which advocated liberty, equality, and fraternity and ended up in a bloodbath called the Terror and a military dictatorship; and the American Revolution which advocated liberty and freedom while the founders were practicing slavery, dispossessing the original inhabitants of the land and withholding voting and basic property rights from females of their own kind. The Americans did write a noble document about freedom, however, which took care of the contradictions by simply not mentioning them. This document stimulated many of the newly independent countries to write their own constitutions. So the concept of democracy as described in the American constitution, was spread widely to the newly emerging or re-emerging nations of the colored men. And when they did not accept democracy on their own, Euromen often forced them to do so. The military government of Japan, backed symbolically by royal tradition, was replaced by the American kind of democracy after the Japanese lost the Great War.

Western democracy probably was the most attractive creed in the world until the Russian Revolution and the spread of Marxism. This Euro-system, with its own paraphernalia, was carried to the world of the colored man somewhat later, though it still became the primary competitor of Western democracy. It had its own sacred document, Das Capital, by Karl Marx.

Much of the paraphernalia of the democratic government process was difficult to transfer to the new nations. Universal suffrage for electing officials was one such difficulty. The leaders of Euro-nations made a great to-do about the critical need to get everybody to vote even though the electorate was frequently lukewarm about participation. Though the Western democracies emphasized that free elections were the base of the democratic way, it was quite apparent that there were many special conditions. Wealth was almost a requirement for being a candidate. And if the runner did not have enough of his own, he promised other wealthy men political plums for electioneering money. There also were ways to manipulate the voters. When the electronic media became fairly well developed in the latter half of the 20th century, a candidate's visual image became critical. It was most natural that one of the popular presidents of the United States had been a movie actor.

Such manipulation was possible even with a population that was mostly literate. So one can imagine how easily voters who were mainly illiterate, as they were in most of the newly emergent nations, could be manipulated. Votes could be bought with a meal or a drink. And most of those who did vote could have little understanding of the candidates because their social status was so different. That countries like India ever got through any elections seems almost miraculous. But the leaders tried, and after a fashion, kept Euro-forms going. Democracy became a sacred word at that time, whether represented by the Euro-Western practice of multiple parties controlled by the wealthy through the media or by the Euro-Easterners in having a single party controlled by an inner circle of persons who had worked their way up in the system.

Politics existed to control the power of a social system. And though such power ideally was passed around through peaceful means, all great civilizations, including that of Euroman, were established and maintained through force. Moreover, the ultimate reservoir of power was always the national army, beginning in the days of the Pharaohs. Euroman was no different. He evolved a military system based on an elite who purportedly took directions from civil leaders and carried them out through the bodies of the lower classes. He developed special codes of behavior, including a clear-cut distinction between the supervisory class and those who had to do the killing. Thus there were special schools to teach the supervisory class (the officers) how to get the killer class (the soldiers) to destroy those on the other side, at first only the combatants, but later on the populations in general. His greatest device for indiscriminately killing off large numbers of people was the airplane and the specialists trained to fly them.

These armies became quite effective, most notably in the colonial wars. So when independence loomed, the new nations did their best to adopt the Euro-system. This included military schools, armies, navies, air forces, artillery, infantry, military discipline, and uniforms. The re-emergent civilizations got so good with Euro-militarism, that they challenged Euroman himself, most notably the Japanese who attacked the Russians and then all the colonial powers in

Asia, and the Chinese and Vietnamese who successfully fought the Americans.

Several aspects of political thought that emanated from Euroman affected the colored men to the day of Takeover. One was something called human rights. Euroman came up with the idea that man had some basic unalterable rights, no matter what his current status. In the Euro-American constitution these were referred to as inalienable, specified as "life, liberty, and the pursuit of happiness." The closest Euroman ever got to defining happiness was that the individual had the right to do what he wanted. It might be worth repeating that these noble pronouncements were first made by men practicing slavery and dispossessing the native inhabitants of their territory. The only way one can understand these pronouncements is that those who made them, like the ancient Greeks, did not really mean everyone. They actually meant well-to-do, white, male landowners.

But through the years many social changes took place, including a very bloody war in the United States which resulted in the abolition of slavery there. It had already been abolished in most other Euro-colonies. But rather than abandon the original document of the slave owners/Indian bashers, the Americans developed a system to modify it which they called amendments. The end product got so complicated, a group of old men had to serve as interpreters of the original document. And by the late 20th century they had reinterpreted the pronouncements of the founding fathers so extensively, that it was difficult to remember that the original founders had been slave owning men in the process of dispossessing Indians, Spaniards, Mexicans and Canadians and who believed that a woman's place was in the home.

It was not so unusual for Euroman to have specialists interpret sacred documents. They had emerged before as interpreters of religious holy books. Both the Jews and Christians had such specialists.

Anyway, the colored men of the world got the later version of Euro-documentation. Human rights must have appeared to many as something that had been there all along, especially if one considered the opinions of the Supreme Court judges. So they tried to copy some of the practices: outright slavery became very unpopular; harsh punishment, including execution, had to be done in secret; deviant sexual practices were to be tolerated; and even women were to be given voting privileges and property rights on occasion. And the consequences throughout the world were many. Slavery was abandoned for the most part, and women got more rights. Perhaps the use of torture to induce confessions, also widely relied on by Euroman through the 17th century, was reduced, though this is debatable. In the old days rather crude methods such as breaking bones, tearing the skin, and immersion in water were often relied upon. But in the new age brought on by Euroman, more advanced techniques could be used. The new torturers were especially fond of applying electricity to the sensitive parts of the body such as the genital or anal area.

It became very popular to espouse human rights politically, and some presidents of the United States were elected with few qualifications other than the ability to make such espousals with an air of sincerity. Human rights became so popular, that righteous organizations like the United Nations and Ford Foundation established special divisions to monitor government activities with indifferent success. By practicing human rights, a government could claim to be civilized. The greatest myth in this respect was promulgated about the ancient Greeks, who were given credit for being the first to practice democracy. In truth, their civilization rested solidly on slavery, class divisions, sexual discrimination, and their own form of colonialism. One of the greatest ironies was that those who espoused human rights loudest, Euromen, were directly responsible for taking rights away from the most people, the tribals and colonized. But as mentioned, consistency was never one of Euroman's strong points.

The human rights advocated by Euroman were those of the individual. Each person was supposed to have been born with them. This resulted in a great dedication to individualism, and in the end, perhaps Euroman's greatest problem. Each person was supposed to be able to move up the ladder of success according to his own abilities. This was basically an anti-social concept, that the group existed only to fulfill the needs of the individual. There were several explanations as to how this dedication to individual needs became so pervasive. But what was more important in the final analysis, these individual rights were not shared to any great extent by the colored man of the world, particularly those who survived the Euro-onslaught. The tribals of the world were more individualistic, but they did not get a chance to reap the benefits.

The nations of Asia, who were fundamentally group oriented, survived best against the onslaught. And though many of their young people flirted with the new concept of individual rights, group forces prevailed in war and peace. And in the end, the greatest challenge to Euro-hegemony came from the socially based power of East Asia. If the Atierrans had not arrived, it seems likely that the individual-oriented civilization of Euroman would have gone down at the hands of the group-oriented Asians. Euroman had become very nervous about the rising power of the Japanese, Chinese, Koreans, Taiwanese, and similar others.

10

Managing the Goods

I concentrated on the line of coins I had arranged on the coffee table. There were only four kinds a penny, nickel, dime, and quarter. The types had steadily decreased in my lifetime until they were now of no value whatsoever. The Atierrans handled all transactions with the personal identity card. They'd had them stamped in the various languages of earthlings. No one had ever seen one that belonged to an Atierran, not that it would matter since no earthling knew their language. The best I could make out was that it was a language of vibrations, something like what symbolic reptiles or bats might have evolved. There were no clearly audible sounds. Anyway, there was so little to buy in their shops that most of us rarely bothered. After all, our food, clothing, and other necessities were provided.

I moved my line of coins in and out, wondering why I had saved them. Most people had thrown them away. They were all as worthless as the penny had become before the Atierrans took over. I wondered if at the fall of other civilizations, say the ancient Persian or Roman, whether the citizens had thrown their coins away. I knew that coins were frequently found by archeologists and collectors in the ruins of ancient cities.

But I had not thrown mine away. I had put them in a small container which I stowed in the single trunk I'd been allowed to take to my new living quarters. And there, after all the months that had passed in the new era, I had taken them out. I realized I was turning into a nostalgia buff, I who had prided myself for long on being an innovator, always ready to try something new.

I thought of how much had changed in money matters even before the arrival of the aliens. I picked up the penny and wondered why I had kept it. The small copper coin had become so useless in the last 40 to 50 years of the Euro era that few persons would bother to pick one up. I remembered that in my boyhood one could buy something real with a penny, a handful of peanuts, a ball of bubble gum, or some candies. Some people were still picking up nickels at the time of Takeover, though I had noticed many in the gutters. But in my boyhood one could still buy an ice cream cone for a nickel, and I think one could pay for a streetcar ride. The silver dime was big money in my early years. One could call someone on the telephone or buy a hamburger with it. And then there was the quarter, one-fourth of a dollar, two bits they called it. Anyway, the twenty-five cent piece had kept some of its value. Just before Takeover one could still buy some food with it, and in Las Vegas one could play the slot machines. There were also quite a few other machines in 2020 A.D. that would give you something for a quarter. You could dry your clothes in a laundromat, about 15 minutes per coin, as best I remember. Even into my sixties I'd stop my bicycle to pick up a quarter when I saw one on the pavement.

I was so absorbed wandering backwards in time that I was not aware that Mary and the lights had come on. When she spoke, I jumped, knocking a coin off the table. It was the quarter; rolling to a stop against the wall. I was getting up to retrieve it when Mary spoke, "I'm sorry, Pete, I didn't mean to frighten you. You were so deeply absorbed."

I got down on one knee to pick up the coin. The replacement hip joints I'd gotten through surgery had alleviated most of the pain, but I still couldn't bend very easily. Ah well, I thought, I should be thankful for the improvement. Going back to sit down, I said, "That's okay, no harm done. I was just ruminating about the reduction in value of money during my lifetime."

"That's the medium mankind used for their exchanges, no?"

"Yes, we were in a money economy as were all the urban cultures of the world. Money was the cement that held their economies together. Although it was an ethnocentric exaggeration, a popular song we had in the seventies was true for national economies. It went, 'Money, money, money, makes the world go around, makes the world go around, makes the world go around, etc.' "

"And now it is all over, and the medium of exchange is the personal identity card, a piece of plastic."

"Well, like all good things, it had to come to an end."

"What were you doing, moving the coins around, Pete? You seemed to be getting a special feeling from handling them. It seemed almost like an erotic caress."

I felt a little like the boy caught with his hand in the cookie jar. But I knew by then that I couldn't get much past Mary. She was like one of her namesakes in the old world, a nurse I had known who had perfected the art of watching people. I said, "I suppose I do. I spent a whole lifetime counting money, and I must say that I got pleasure from it on occasion. So now that I'm in a retrospective mood for the project I've also become retrospective about some of my own experiences."

"Like what regarding money?"

"Well, there I was studying the few pieces I still have and I couldn't help thinking how they had changed long before the arrival of the Atierrans. Some had even disappeared. When I was a kid, and into my early manhood, we had a silver dollar and also a half dollar. These were taken out of circulation for reasons deemed appropriate by money specialists. But the denominations that remained, including paper money, became much less valuable in my lifetime."

"You mean the substances they were made of decreased in value, copper for instance?"

"Oh no, the basic mineral value of a coin was not what gave it purchasing power. Even gold and silver, which were thought of as the premier minerals, went up and down in purchasing power. Many factors affected the real buying power of money. And I, like most others, could hardly understand them. We had a kind of social scientist, called an economist, who explained such events as inflation, deflation, adflation, superflation, and other esoteric money fluctuations, but most of those gurus did not have a very good track record. For instance, we used to have a periodic economic event called a stock market crash which usually occurred as unexpectedly as an earthquake. You could get any explanation you wanted for why it occurred if you found the right economist. In accuracy of prediction they were about as good as our psychiatrists in the medical field."

"You say these economists were social scientists. Aren't you anthropologists also supposed to be social scientists?"

"Yes, we have some things in common, hard as it is to believe sometimes. Many earthlings still did not think of economists as social scientists.

"Our investment system, called the stock market, could go up or down or become bearish or bullish or do all kinds of things on its own, as if people were not manipulating it. The way the stock market was explained by economists, it was hard to believe that people were causing things to happen. In actuality, money lost or gained value because people raised or lowered prices or because other nations would only exchange their money for yours at a lower rate than you expected. Someone in the other nation made a decision to do so and this was implemented by others in the government.

"The alternate political system I discussed before, Communism, differed in its economic policy. Their gurus figured out a way to avoid the ups and downs of money value, which was to refuse to exchange their money with other nations except at their own arbitrary rates. So when they needed to buy or sell something to another nation, they either used gold or exchanged goods through barter, both older systems than stock markets and money marts. Of course, there was constant argument between advocates of each system as to which was better. I might say that *sapiens* tended to argue most about matters he understood least.

"We used to have a saying, 'Never argue about religion and politics.' Both were belief and behavior systems about which little was understood. I think it would have been appropriate to have added 'economics' to those two."

"Anyway, anthropologists treated an economy as a form of social organization, a system of group behavior for the production, distribution, and consumption of the products of a people's technology. This view was more apparent in the simpler societies which did not have markets or money, but still managed to process their goods. But the economic activities of urban industrial man still constituted social behavior."

"Many people found economic discussions very boring, if not incomprehensible. And I did also until I learned the anthropological way of explaining the system. Of course, I could get by with understanding simple systems since those were the kind that anthropologists ordinarily dealt with. Perhaps this tells us something about a main characteristic of anthropologists. They were simple people."

"Ah ha. You are now telling me that this part of culture, what you call the economy, was a part of a people's social system like political organization?"

"That's it, the social relations of an economy were very important, second perhaps to the social relations of power control. Even before market economies evolved."

"So that I take it, is where you want to start, in the premarket economies, right?"

"Right, because just like the organization of social power, the organization of economic relations throughout human prehistory was relatively simple. And these were the ones anthropologists studied. For the most part, the economics of complex societies were left to the gurus of finance."

"So, how did mankind manage through most of his existence as a biped?"

"Well, the simpler societies, bands and tribes, generally produced only what was needed immediately, and goods of limited variety. Not only that, there was little specialization of labor except between men and women. And generally they complemented one another rather than competing. That is, the work men did, like hunting, was complemented by the women's work of making clothing from the skins and cooking the meat, as well as gathering seeds and berries. The foods and other goods were then distributed to family members, fellow kinsmen, and neighbors. It was a distribution system that had no need for money or markets, called reciprocity by anthropologists. One was supposed to be generous in that kind of society. Excess goods and money were not squirreled away as was done in complex economies."

"Consumption standards were more or less the same for everyone. There was a limited variety of goods and no real luxury items. Everyone ate and used the same kind of things. A very good hunter or gatherer might have a little more meat or nuts than the neighbors, but he/she was supposed to be generous, which tended to even out the amount consumed by an individual."

"As cultures became more complex, particularly in their technology, there were more surpluses which made some trade possible. Lacking true markets or money, however, these tribal people relied on simpler procedures. They exchanged goods through barter. Thus the growers would trade bananas for the hunters' elephant meat. And when societies got even more complex, men worked out systems of redistribution. The "big men" or organizers would arrange for their followers to bring in their yams or pigs or smoked fish or fish oil, for a feast at some ritual event. Much prestige would be derived in the process. Anthropologists have described this big man pattern of redistribution in excruciating detail, particularly for New Guinea and the northwest coast of North America. The economic gurus of high finance were hardly interested in such goings-on of primitive people. They, like most other academics who studied mankind, began with the ancient Greeks."

"All this began to change rapidly about ten thousand years ago when technologies and social systems became much more complex. The most significant changes resulted from the domestication of plants and animals. This soon permitted a surplus. Since people could grow more wheat and cattle than they needed for their own use, they began to see the value of true exchange, offering some of their excess wheat or cattle for the lumber or copper of their neighbors, or even distant tribesmen. Market systems evolved which soon produced a need for a common medium of exchange, money."

"These developments led to the rise of cities where economic changes occurred at a still more rapid rate. And as exchange became more important, it became worthwhile for the craftsman to specialize. A man who did nothing but make gold ornaments got very good at it and could earn a lot more than a jack-of-all-trades. But the goldsmith depended on the market absolutely, not only to sell his ornaments, but to get his food, clothing and all else he needed. Thus, such specialization of labor depended primarily on money, markets, and city living."

"There were consequences and one of the most significant was the constantly increasing variety of goods that was bought and sold. Very soon some people became richer than others. Many techniques for earning money evolved and the most innovative of all was to manipulate money to make more money, without bothering to produce anything. Anyway, those who got richer began to consume the better kinds of goods and more of them, and to use those who were less well off to become richer still. Thus while the poor got by with boiled wheat or rice and a little chili to make it palatable, the rich had a variety of meats, fruits, and vegetables as well as the wheat or rice. All of a sudden it became possible to get fat. Prior to city life, few people could afford to overeat. The same was true of other commodities. While the poor wore the plainest cotton garments, the rich wore linen and silk. Classes and castes, slavery, peonage and other methods for exploiting workers proliferated. And because there was a surplus, it became worthwhile to go to war to take the goods of others. And the most efficient way was to turn workers into armies that could fight the workers in other armies, allowing the winning side to take the goods of the losers. Warfare for loot lasted throughout the pre-expansionist period of Euroman."

"And with some minor variations this was the economic world of Euroman when he departed from his shores."

Naturally Euroman brought with him a number of ideas which had emanated from his own economic system. One was his notion of property. All peoples seem to have had ideas of "mine" and "theirs," but in simpler societies the idea of exclusive, individual ownership was not as common. In general, band or tribal people felt they were the owners of the tools of their trade which usually were fairly simple. Thus the hunter claimed his weapons while the gatherer claimed her baskets. But land was not usually claimed individually since in the nomadic life of the hunter-gatherer, mobility was more valuable than claim to a small plot. The whole band or tribe claimed the entire territory.

And along came Euroman who had been raised with the idea that property, including land, should be individually owned. He also lived in a society with many social levels, the top of which was the royalty. The king and queen could claim any new territory that their subjects could conquer, which they would then parcel out in royal grants. And that is what happened in one land grab after another. Hernan or Fidencio or John or Peter claimed territories all over the world in the name of King Ferdinand and Queen Isabella or Queen Victoria or King George. Wherever they landed, the Euromen would stake claim to enormous areas about which they knew practically nothing. Some time after the claim was made, an expedition would be sent forth, the natives smitten, and the conquerors would establish themselves as the new owners. This was the usual pattern of takeover in band or tribal territory, though the Spanish did the same to the two major civilizations they conquered in Meso-America.

When land was claimed in the name of the king, the land grants would be made to the loyal conquerors and their descendants. In California enormous grants to individual Spaniards were recorded in the history books. The land had of course been occupied by Indians before the Spaniards came. When these southern Euromen lost the war to northern Euromen, the Americans, the land again was taken over and parcelled out among the new conquerors. The same system applied in eastern North America. George Washington was given thousands of acres of Indian land for his service in the French and Indian War. This was in accord with Euro-standards. It is no accident that the father of Euro-America was a land surveyor.

The old urban civilizations of Asia posed more serious problems to Euroman. For one thing, they also had the idea of private land ownership and in addition, they had large populations and complex societies. Euroman must have seen quickly that marching onto the shore of Japan or China, planting the flag, and claiming everything as far as the next ocean or border was rather presumptuous. So he did what was next best. He established enclaves on the shores, primarily for trade, and tried to dominate the hinterland from there. Some of these trading centers became major cities: Bombay, Madras, Calcutta, Singapore, and Hong Kong. All of these and more have, of course, reverted to the original nations after the decline of Euro-power.

The other commodity of value that Euroman was interested in taking was people, and the primary way of doing this was to enslave them. Euroman already had a system for exploiting the lower classes at home. It was called serfdom. But with so many "bodies" available in the new territories, he decided to be more direct; he captured them by force or bought them from other natives and transported them to places where they could be made to work. This system was called chattel slavery under which Euroman had full rights over his commodity. Although this was instituted earlier by the Portuguese and Spanish, it came to full development in the hands of the British and Americans. The primary region where slaves were acquired was Africa, and the primary function was as plantation labor, with cotton and sugar as the main crops.

Later on, when slavery was abolished, the British devised a substitute system, indentured labor, with workers that were primarily from India.

The long-range consequence of slavery, indentured labor, and the form of serfdom that evolved in Spanish areas was the establishment of a class/caste system with the agricultural laborers at the bottom as usual. Even when slavery was abolished and Indian workers in Latin America were given some civil rights, the class system remained with only slight modifications. In the U.S. until Takeover, the poorest classes continued to be the descendents of African slaves and the descendants of the dispossessed, the Indians..

Euroman took money and markets to wherever he went. Where these already existed, as in Asia, he picked up some additional ideas. The concept of cash and paper money came from China. In Asia and the Middle East Euroman was then able to sell his own commodities and have others made locally for his own use and for trade. The main such trading area was East Asia where there were large populations. At first he dealt in trinkets, but later he got tea from China and opium from India and Turkey which he peddled in other areas. For example, he sold the tea to India, and the opium to China. Then he dumped cloth made from cotton grown with slave labor in the Caribbean and manufactured by factory laborers in England.

Since money and markets were totally new to the tribal areas, to introduce them, Euromen devised a new institution, the trading post. This was a store where Euroman could trade his products for the local commodities. The most extensive networks of trading posts were established in North America, where beads, glassware, copper and iron utensils, knives, guns, and hard candy were traded for wild animal skins, particularly beaver. In the Far North the posts traded in skins of the Arctic fox, and along the Pacific Coast in sea otter skins. Beaver skin was in great demand by Euroman throughout the 19th century, used to make hats for men and coats for women.

Trading posts brought great changes in the Indian way of life. The Indians first learned to barter and then to use money. The attraction to new products was so great, Indians quickly abandoned their former way of life. Why bother making a bow and arrows when a rifle was available. What Indian wife would want to bother cooking in a hide or pottery vessel when an iron or copper pot could be had. In addition, Indian women who had made decorations out of porcupine quills which could be threaded and sewn on buckskin, could substitute glass beads, which were shinier and more colorful. Thus, beadwork became the "traditional" Indian decoration. Satin or sateen also was cut into pieces and sewn on garments.

Needless to say, the Indian trapper was in deep trouble once the wild animals were hunted out and/or fashions changed among the Palefaces. But by that time it was too late to recover the old ways. He had been drawn inextricably into the Euro-market network.

A particularly wild native spree brought on by new trading patterns had great fascination for anthropologists. It was called a "potlatch," and was practiced by Indians along the North Pacific coast of North America. The classic description was of Indians assembled at great feasts, competing with one another in giving away or destroying commodities. One popular trade item was the Hudson Bay blanket which they would hurl onto great bonfires to show off their wealth as compared to their rivals. In the early reports, anthropologists thought this was a "normal" event though even then the extent of property destruction seemed a little overdone. But in those days cultural relativism was in. That meant if natives did something, it was all right. How could anthropologists criticize the source of their bread and butter? But anthropologists of the next generation, when classic cultural relativism was fading, came to the conclusion that the great destruction was a recent cultural aberration, a consequence of the new wealth that had come to these Indians from participation in the fur trade. The Northwest Coast Indians were especially rich in fur animals, especially the beaver and sea otter.

Toward the end of the great trapping orgy, the maverick Euroman, the mountain man, joined the Indian trapper. But his place in history was brief since he soon joined up with the incoming or Euro-settlers. They, of course, after initially using him as a guide, had no use for him or anyone else whose way of life consisted of wandering around the country and mucking around with Indians. The traditional trading post, based on money transactions, remained a viable institution on the Indian reservations of North America, serving social needs as well as being an outlet for Indian crafts. One of the popular latter day commodities was soda pop since Euroman in his wildly contradictory policies toward Indians, forbid the use of alcohol on the little enclaves. On a typical trading post in the late 20th century a group of Indians could be seen lounging about outside sipping canned pop. Their brothers were at liquor stores just outside the reservations where they had gone in their pick-ups for booze.

The most extensive networks of trading posts couldn't continue as such when the fur trade died out. Probably the biggest in that industry in the United States was the American Fur Co. which made the fortune of John Jacob Astor and his relatives. I believe they simply diversified the business when it was no longer profitable to deal in beaver skins. However, the great Canadian fur traders, the Hudson's Bay Company, after dominating the trade with Indian trappers from the Arctic to the U.S. border for 200 years, gradually changed to become one of Canada's biggest department store chains. It kept its old name and even continued to sell its highly touted woolen blanket.

One overall economic trend with all its trappings, disseminated by Euroman, was capitalism, the system of investing money in various ventures to produce items which could then be sold, thus making money work to make more money. Western Euroman became very proud of his achievements as a capitalist. One economic guru even suggested that it derived from the reformed version of his religion, Protestantism, calling it the "Protestant ethic." Anyway, Euroman invested wherever he could get a foothold and through the 19th century did very well. The British were the supreme capitalists until the early part of the 20th century when they were replaced by their cultural descendants, the Americans. By the end of the century, however, the center of capitalism had shifted to East Asia where there were few Protestants indeed. These were the cultures which had survived the Euro-shock best by using Euro-methods against Euroman. One such was capitalism. The country which did particularly well was Japan which took over many Euro-customs. They industrialized early and built a Euro-style war machine after being forcibly "opened up" by Euro-Americans.

A part of the capitalist belief system was that there should be free world trade. This was advantageous to Euro-man in the early days because he was the primary producer. The Europroducers needed markets so the idea that a large population could be off limits, as in pre-Meiji Japan was anathema. In that instance, the new industrialists, the Americans, sent a military mission with an ultimatum to the Japanese who were then trying to keep Euroman off their backs by closing the borders. Since they were outgunned, they opened their doors.

But then instead of becoming a passive market, they decided to get into the race. They industrialized and engaged in free trade. But the complex Japanese culture had other tendencies, one of which was a martial spirit. The Euro-pattern of colonial domination appealed to a certain element, and they launched their own aggression which they euphemistically called the East Asian Co-Prosperity Sphere. Of course, just as Euroman had done before, they were really thinking about their own prosperity. If Euroman had considered the Japanese as his true cultural heirs, he would have been proud. Unfortunately for him, they became too good at his game.

But unfortunately for the Japanese, Euroman, mainly the Americans, still had considerable industrial and economic power, and the Japanese lost the war.

And of course since ethnocentrism prevailed, the Americans imposed their own form of government on the Japanese, including a restoration of capitalism which the Japanese had chosen themselves in the mid-19th century. And so after a decade or so of tightened belts, the Japanese, like the other losers of the war, the Germans, emerged as super-capitalist producers. The British had considered the pound sterling to be the foundation of international currency during their heyday. By World War II it had become the U.S. dollar, and by the 1970s it was the Japanese yen and German mark alongside the dollar.

So, the "push-pull" effect resulted in capitalism being adopted in most places in the world. And along with it went national and international banking systems, common stock exchanges, corporations, floating interest rates, and other financial machinations that had evolved in Euroman's world.

Capitalism was so widespread at the end of the 20th century that many people thought it was inevitable. It is interesting to speculate about other possibilities. One was the traditional Islamic system which condemned loaning money for interest. Usury was a vile sin as it had been in classic Christianity. According to the mythological history of Christians, the Bible, the great savior, Jesus, got angrily excited and threw the moneylenders out of the temple. And, of course, another noncapitalistic approach was Communism which condemned the Western system as being exploitative, which probably was true.

11

The Local Group

I sat quietly, listening to voices from the past. In my fantasy a child said, "Just one story about Turkey-Lurkey, Pete."

I heard myself say, "I'm tired kiddo. Couldn't we do two tomorrow?"

I had been momentarily taken by the child's form of address. There I was, acting like a father while he was addressing me by my given name. Laura and I had agreed, largely at my suggestion, that we would have our son address us by our given names rather than as Mom and Dad. It was only later that I realized this had been a small protest against parenthood. I had not really intended to be a father. Little did I know then that I was in the forefront of an ongoing social process, the weakening of the family, with many of its functions being replaced by special interest groups. The old forms of address which denoted one's relationship to other kinsmen were becoming less important, as were kinship roles.

Harold was not a whiny kid even though he was normally self-centered. He said, "I was counting on it, Pete. I really like those stories."

I tried to think how the story telling had ever gotten started. The best I could remember, I had gone in to talk to him one night after Laura had put him to bed, and neither he nor I could think of anything that had happened recently that was of interest to the other. So, after a not very great conversation, he'd said,

"Why don't you tell me a story, Pete?"

Casting around for something to keep us going for ten or fifteen minutes, I asked, "What about?"

"Oh anything. You always tell good, real-life stories, especially about the animals you had when you were little. Why don't you tell me a made-up story about one of them? You could even make up the animal, too."

Amazing, I thought, how such things got started. A man in his early fatherhood and a seven-year-old boy trying to work out a system to while away a few minutes of togetherness so the father could feel he was fulfilling his responsibility and the boy could go to sleep. I said, "Oh, I don't know, Harold, making up a story is quite different from telling one that actually happened. It takes a lot of imagination."

Since his eyes were getting brighter as he got more enthusiastic, it seemed to me that this was not a great way to put a kid to sleep. But he wouldn't quit. "I know, Pete. But I wish you would try."

He hesitated, then picked up another idea. The kid certainly didn't lack for ideas. I was stroking his soft, blond hair when he said, "Pete, why don't you light up your pipe. That'll help you think of some animal and how to get started."

"Okay kiddo. Give me a minute."

I took the pipe out of the side pocket of my jacket and the pouch of tobacco from my hip pocket. Laura had given me the pouch for Xmas while Harold, with her help I'm sure, had given me the pipe. They evidently both liked me to smoke a pipe. I couldn't help fingering the piece of charred pocket lining as I took the pipe out. It was a problem of mine, putting a still smoldering pipe into my pocket. Laura had said several times, "You will burn yourself up someday."

Anyway, after lighting up and taking a good draw, I said to Harold, "Okay, I've got it. I'll tell you a story about a turkey, Turkey-Lurkey I'll call him."

The kid's eyes lit up even more. Little did I know what I had started.

He said, "Oh I like that. But why did you decide on a turkey?"

I thought for a bit, but couldn't come up with a reasonable explanation. It was a couple of weeks after Thanksgiving when we had roasted a turkey. But apart from eating them on festive occasions, I had never had anything to do with turkeys. I had to say, "I really don't know, though it is an important bird symbolically to Americans. It was supposed to have saved our ancestors from starvation when they were trying to get a foothold in Massachusetts." I threw in, "You like to eat turkey, don't you?"

He quickly went on, "Turkey's okay Pete. Why don't you begin?" I suspect ne said that because he was afraid I was going to give him a lecture. Even then I

135

was at the point where I would lecture practically anyone if I got the slightest chance.

And so I began. "Once upon a time there was a turkey that lived in a big valley where there was plenty of food. But one thing was missing, other creatures. Not only were there no other turkeys, there weren't even any other birds, or animals, or people. So one day our friend, who was named Turkey-Lurkey, took off. He headed out of the valley."

At that moment I realized that Mary's screen was lit. I read, "Monitor."

It startled me. My mind had drifted so far away.

Mary said, "Hello Pete, how are you?"

"I'm okay. And you?" I still couldn't help reciprocating in human fashion even though I knew that a computer was an unlikely receiver.

"Oh, you know we computers rarely have anything go wrong. What fantasy were you in when I turned on? Your lips were moving."

I laughed, slightly embarrassed. "Oh Mary, you'd never believe it. I was reliving in memory an experience of my early manhood. It's one of the prices one pays, I suppose, for living to be so old."

"That's not so bad when you consider the alternative, is it? At least, you had some interesting things happen in your lifetime. And now you've got me intrigued. What were you saying when you were moving your lips?"

I felt that I was blushing, and that embarrassed me also. It seemed inappropriate for an old fellow like me. I decided to tell her. She always seemed to be able to worm things out of me sooner or later anyway. I said, "I was retelling a story I used to tell my son when he was little."

The more I thought of it, the sillier the explanation sounded, but I went on. "The story didn't really make any sense. I just made up this character, Turkey-Lurkey, and night after night I made up adventures about him. Since it was just a little fantasy to help my son go to sleep, I simply told him anything that came to my mind." I stopped, caught up in my reveries again.

"Pete," Mary said softly. "You're drifting off again. Why don't you tell me what you're thinking? It might help clear up your own thoughts."

I shook my head to get the images out, amazed at the new thought that was entering my mind. Mary was turning into a therapist. My computer, my shrink, I thought, chuckling inwardly. But it wasn't such a bad role for her. I said, "I was wondering why I did it, you know as a social role."

"You mean why you felt responsibility as a father?"

"Yes, but more than that. You know, the last couple of sessions we've been talking about social roles rather than individual behavior. The social scientists developed the concept that apart from particular characteristics, each person had a number of social roles, either achieved or ascribed, as we used to say. One was a man or a woman, an ascribed role; and one was a lawyer or shopkeeper, an achieved role Well, the social scientists decided that each person had a number of such roles, the greater the complexity of the society, the more roles. And for each role there was some expected behavior. For instance, in my early middle manhood, I had the roles of father, husband, son, brother, professor, middle-class white, and American. I was not really a joiner, so I had fewer roles than most people. As members of organizations, churches, political parties, bowling clubs, unions, they took on additional roles.

"I see. But what has this to do with your account of Turkey-Lurkey?"

I grinned inwardly, recognizing this as one of my old professional tricks — to follow tortuous paths of ideas in order to get across seemingly unrelated events or conditions. It was a practice brought to the ultimate on the media by another professorial type, James Burke.

I answered Mary in my best professorese, "Okay Mary, this is how it goes. I've already discussed two bases for group membership, power control and economics. One was a member of an organized unit, say a nation or a political party, and one was a member of an economic entity, say the upper or lower class. Each group had certain expected behaviors. Lower class people in America drove used American or cheap foreign cars while upper class people drove Cadillacs or Mercedes. The poor ate hamburger, the rich ate filet mignon.

"But there were other bases for group membership. Two were of great interest to most people. The first was based on kinship and the second on what social scientists called special interests. Kinship existed from the recognition by all peoples of what they called blood relationships, although in truth blood had nothing to do with it. People were kinsmen because they shared genes, not blood. Kinship was a way of grouping people together because they had a real or supposed genetic relationship. The basis was the nuclear family of father, mother and children. People were grouped together by special interest because they had some common goal."

"All this sounds pretty theoretical, Pete. And though I recognize the need for theory, I'm also a stickler for specifics as well as charged to keep to the subject at hand. And you have already admitted that one of your weaknesses in teaching had been your tendency to wander off on tangents."

"That's true, though the connection between theory and specific explanation usually seemed clearer to me than it did to others."

"That is probably the usual case. But I'm afraid I will have to bring you back to Turkey-Lurkey if you wander too far. What is the connection between your telling that story to your son and this business of social roles?"

I felt almost gleeful as I launched into the clincher. I always had got a feeling of accomplishment when I came to the final explanation, the one that tied everything together. I said, "Okay, here it is. The local group that was most important to most people in most societies was the family. Further, depending on the type of family, there were a number of expected behaviors for each type of individual. Being a member of a family, I was like most others in this respect. And being an adult male, I was the father. Further, I was living in a social class which expected its fathers to take an active part in raising their children. Moreover, they were pushing permissiveness in those days. A whole generation of children were produced which we came to call "rotten kids," children who were encouraged to do their own thing by their middle-class parents. When they grew up, many became Hippies and "flower children" who condemned the way of life of their parents because it was so materialistic.

Ideally, fathers in those days were supposed to be pals of a sort to their children much more than disciplinarians or guides. And though I was far from being the ideal permissive father, I had gone in that direction. And there I was helping put my son to sleep by making up a story for him. My wife, Laura, would take a complementary role on alternate nights by reading to him from children's books."

"So, you were reliving that experience of your early parenthood just now?"

"Right. And I suppose it's because of what we've been talking about the last couple of sessions, social organization or being together. My mind had drifted to the last important kinds of groups I knew of, those based on kinship and special interest. And as usual, I thought of one of my own experiences."

"Are you saying that this will complete your discussion of social matters?"

"More or less. After going through what mankind generally, and Euroman particularly, did about these matters, we can go on to the final topics. They are becoming fewer."

"All right, let's do it. First, how did mankind manage through most of his history with the local group?"

"Well, again, we had no direct information of the earliest days since social behaviors were not what lasted over time, grist for archeologists. So as with most other cultural behaviors, we anthropologists turned to the living primitive peoples for information, assuming that what contemporary hunters and gatherers did was probably roughly similar to what our distant ancestors had done. This certainly wasn't easy because there weren't many primitives left when anthropologists came on the scene, and those that remained were going fast.

There were none left by the end of the 20th century."

"And so how did those primitives manage?"

"Mainly they depended on kinship. They were family people where the father and mother took complementary roles raising their children in the hope that they would turn out like them. This business of the father helping raise the children was a very human action. Except for the apes of Gibralter, and a small South American monkey, the other primates left it to the mother.

"The nuclear family was basic to all the extensions of kinship, though some groups got it mixed up with larger groups like clans."

"And other basic units, those you call special interest groups, when did they come on the scene?"

"All we know about their origins is that there weren't many in the simple societies. But still one could join a curing society or a soldier society. Basically however, since there were few social choices, there was little need for special interest groups. For instance, in the typical primitive society there was only one set of religious beliefs, leaving no need for a variety of churches. This situation was quite unlike that of Euroman. He had so many different religious systems, one had little idea where another person stood until he told you what church he belonged to."

"So, did all this change when the great cultural revolutions took place, those you called domestication and urbanization?"

"Sure, because whenever societies became more complex, new social roles evolved. The great city, even the ancient one, already had a number of religious choices. In ancient Egypt, for instance, one might be a follower of the cult of the dead, a worshipper of Akhnaton's sun god, a fire worshipper from Persia, or even a Jewish monotheist. Membership in any one of these groups would produce different behaviors. The same would be true if one were a craftsman and member of an early trade union. As one moved toward modern times, there was decreased dependence on kinship to arrange one's affairs and increased dependence on special interest groups. And so it was for Euroman when he started on his voyages of discovery."

Even if family relations had eroded to a certain extent through the history of man, Euroman still believed strongly in them. Further, the nature of the family continued to change during the period of Euroman's adventures in the realms of the colored man.

In 1492 the Euro-family was basically patriarchal and authoritarian. Woman's place was secondary and in the home, performing the duties advocated much later by the Nazis of Germany: managing the cooking, cleaning, children and church. Her political and economic rights were few.

Marriage was a social and religious ritual to be sanctified by God and society. The union was considered to be an affair of reproductive responsibilities rather than one of sexual or other pleasure. Sex in the family was considered to be appropriate only for the purpose of producing children.

One was supposed to have one spouse only. Incest taboos prohibited marriage between close relatives on both sides. And though the selection of spouses was still strongly influenced by parents, the idea of romance as a reason for mating and marriage already existed, at first only in the upper classes.

Divorce was strictly frowned upon as contrary to the laws of god. Although Euroman used "the laws of god" as a justification for all kinds of behavior, he particularly relied on these for governing sexual and family matters. Sexual morality was considered to be a matter in which the Christian god was very interested.

Old age was considered a family matter, older people being respected for their experience. They were cared for in the family residence.

During the next 450 years, Euroman's ideas and practices changed much in regard to the family. About the only customs that remained basically the same were the incest taboos and the monogamous nature of marriage. All the others changed considerably as the Euro-economy changed. The role of the father steadily declined until by the mid-20th century, the ideal marriage was considered to be a partnership. It never was exactly, though women's rights became a big issue and women did get quite a few prerogatives which had been unthinkable in the earlier era. So far as the children were concerned, the mother's role declined also so that by the mid-20th century permissiveness and children's rights had become widespread. Children were thought of simply as little adults rather than inexperienced, unformed pre-adults. Monogamy remained the standard, though marriage itself became less attractive. While at the end of the 15th century non-marriage, except for clerics, was unthinkable, by the end of the 20th a large proportion of men and women did remain single. Couples developed a pattern of "living together," without formal social validation. There was an increase in sexual dalliance, or at least talk about it. Men would have sexual affairs on the side while remaining married. Incest taboos remained strong. Sex and marriage were forbidden between persons related as close as cousins.

One of the most far-reaching changes was the steady increase in the belief in romance (love) as the prime justification for mating and marriage. By the early 20th century one was supposed to "fall in love" before getting married. This was thought of as an emotional state bordering on mania which an individual was powerless to resist. Love became the chief plot device of books, films, television, and video. However, the emotion was more feminine than masculine. While men continued to think of their partners as sex objects, women thought of theirs as soul mates. The economic and religious bases for marriage were pushed into the background, when not ignored completely. If there was love, there was supposed to be sex for fun. Sexual modesty decreased, and something called a sexual revolution took place. This included the idea that people should be allowed to do what they wanted sexually so long as no one else was bothered. In fact, the actual changes in sexual behavior were probably not as great as they were thought to be, but people did talk about sex a lot more openly. While in the beginning of the 20th century adults would not use words that referred to the sex organs, by the late 20th century liberated females felt it was okay to use four-letter words. These came from Anglo-Saxon and were considered to be more direct and risque than multi-syllabic words which came from Latin but meant the same body parts or functions.

One behavioral change that did take place however, was that homosexuality became permissible, even if it was still viewed askance by most people. Undoubtedly homosexuality had been practiced in Euro-society for a long time, but had been condemned so strongly by the Christian church, it had to be done in secret. In the 20th century, homosexuals came "out of the closet" and took other steps to make themselves legitimate. Among other things, they rejected the old term for themselves and began calling themselves "gay," though there was never any indication that they were any gayer than anyone else.

Humankind, and particularly the English-speaking peoples, had long had the practice of adopting new terms once old ones had become overloaded with negative connotations. The very dark men of Africa who had been brought to the New World as slaves tried desperately to change their social status by adopting new words for themselves among other things. In my lifetime they went from nigger to negro to colored to black. And as mentioned, we had a double set of terms for body parts and functions, those derived from Latin for ladies and during the "modesty" era of sex, while those for men, the four letter words, came from Anglo-Saxon. During the period of sexual openness "forward" ladies used . the four letter words also.

But back to the rest of the changes in family matters. Since sex and marriage were considered to be for fun, when they no longer served that purpose, Euro-couples opted for divorce. It got to the point where no cause was required, but the implication always was that at least one of the parties was not having fun. Social and religious reasons were hardly considered, at least in civil law which is where divorces were granted. Thus, in the late 20th century divorce and subsequent remarriage became a frequent occurrence.

Older people often did not remain with the family. Not only were they not looked up to for their experience, they were usually looked down upon as old-fashioned. Frequently used terms for older family members were "pops" and "gramp," both with negative connotations. Older people, called senior citizens, often lived in communities exclusively for the aged and when they became decrepit, they were placed in what where euphemistically called retirement homes. They, and the young people who put them there, knew that such homes were usually their final residence. When they did leave, it was usually in a hearse.

How did all this affect the colored people of the world? In the first place, the effect was less significant than were other social matters, since what goes on in a family is generally more private than what goes on in public, such as the economic or political arenas. Even so, Euroman tried to change people to his way whenever he could. Most Euromen were convinced that their ways were best. So the colored men of the world had to deal with the differences between their own system and that of Euroman even while it was steadily changing.

Probably the most influence was felt during the early period of Euroman's intrusion when the patriarchal, restrictive type family was standard. However, the influence continued even up to the bilateral, open-family system of the 20th century. By then Euroman had swamped the world with films, television shows and videos. These delivered the message that true love could be found if one looked hard for it, and that marriage was based on this sublime, maddening emotion, enabling one to live happily ever after. Also portrayed was that Mom and Dad had little control over their children and that parents were equal partners. Finally, the message was that this kind of relationship frequently led to divorce.

The colored man had to figure out which message was correct. He had learned early that all Euromen were not the same, particularly in sexual matters. Most of the first intruders were men of action. Usually there had been nothing in their upbringing that would keep them from satisfying their sexual drives with persons of other cultures, and to whom they were rarely accountable. So when they had access to dusky, nubile maidens, they ordinarily went at it. They must have seemed very sensual. How could the bare-breasted native ladies then know that these same men frequently lived lives of sexual repression, or at least had strongly curtailed behavior, in their home countries.

Once the conquerors had pacified the natives, along came their brothers, the clerics, and later on, the teachers, who carried the message that fornication and anything that would lead to it was sinful. The missionaries were particularly uptight about exposure of female breasts, lewd dancing and singing, and saw to it that these were strongly repressed.

The exposed female breast still titillated Euroman toward the end of the 20th century, so much so that many males would go some distance to view women at topless beaches. Euroman also commercialized this suppressed desire in the age of the sexual revolution by opening topless bars and restaurants.

Anyway, the colored received these mixed messages about Euroman's attitude toward sexuality in the early days but overall, the missionary attitude was the more influential. People of cultures that had been relatively open about sex and the body learned to curtail or hide their activities and parts.

And what about the other characteristics of mating and the family? As with almost all other customs, Euroman reacted quickly whenever he encountered those which were different and went against his deeply held beliefs. One was the number of wives a person could have. Though Euroman's Old Testament prophets had indulged in the widespread practice of polygamy, some in the mythological account having hundreds of wives, the New Testament did away with that. The Euro-male first met by the colored person had only one wife, even if she wasn't with him. Moreover, the missionaries condemned the practice strongly, and as was their custom, they punished the transgressors, even refusing to let them worship in their churches. Thus, all over the world, polygamy died out or was greatly decreased.

The one great holdout was Islam because their holy book said that polygamy was all right. Of course, the words of Allah as presented by Mohammed do seem surprisingly like those of the Old Testament prophets. But, in any event, once a custom was written in a sacred book, the followers became very self-righteous about it, no matter what religion. The upshot was that even up to Takeover, some Muslims continued to openly practice polygamy. But even in their world it decreased because Euroman had gotten across the idea that having only one wife was more civilized than having many.

And what about love and romance as a basis for marriage? A popular song in the '60s proclaimed that "love and marriage went together like a horse and carriage." Much of the romantic sentiment in Euro-culture came from such songs.

In most of the world of the colored, no matter how emotional a male and female might get about one another, marriage was a social business, and not to be left in the hands of inexperienced young people who were under the influence of their sexual desires. In many cultures, and particularly those with strong rules concerning inheritance, most marriages were arranged by the elders, and when not, their approval had to be obtained. Even Euroman in the 15th century considered that someone of the "older and wiser" generation had to approve marital choices. But this changed drastically over the next 400 years. Eventually one didn't even have to elope in order to marry against parents' wishes.

This version of love and marriage reached the colored late in Euroman's reign, first through his literature, particularly novels and poetry, and later in his films and the electronic media. And wherever traditional society weakened, some young people went the way of romance.

Even so, the older generation in the strongly social societies still had considerable authority over newlyweds if they planned to stay in their home territory.

India had a long tradition of arranged marriages which was still prevalent in the '50s when I did my first field work there. We anthropologists had to have firsthand experience in another culture in order to get our union card in those days. Anyway, many of those I interviewed were men who had come to the city to find work. They invariably insisted that the only way they could get married to a "decent" woman was to let their parents or other senior relatives arrange it for them. Men who tried to find a spouse on their own would be considered mavericks, if not criminals, and no "decent" girl would marry them.

I also got to know a professional couple who had married without their parents' blessing. He was a doctor and a Christian, she was a high-level radio specialist person and a Hindu. The parents did what parents do in non-permissive societies, they stopped speaking to their children. Ostracism is a powerful social weapon.

Many young men and women in India "fell in love," usually with socially inappropriate persons, but few married those persons because of the social penalties.

The exceptions were the countries that adopted Marxism. Karl Marx, a European, advanced the idea that women should be liberated from the shackles of the patriarchal family. So where his views were adopted, women were given new rights and then transformed into full-time workers. According to Marx, both the economy and women would benefit. The traditional family came under attack and marital partners were allowed to select one another. "Being in love" was made more acceptable, though social and economic considerations were not abandoned to the extent they were in the Euro-world.

Euroman concerned himself with woman's position in the lands of the colored where he dominated or was a strong influence. He came down on practices that violated his idea of propriety. Thus, wherever women were being mistreated (according to his standards), he applied civil laws with police force to enforce them. There was no hesitation with the tribals. They were simply stopped from continuing such practices as cutting off the nose or severing the tendons of an unfaithful wife. Even such a practice as cutting off one's finger when in mourning for the death of an important family member was stopped. The English, whose control of India was considered to have been far from advantageous for the Indians, still felt righteous about what they did. They condemned and actively sought to stop all practices which violated their moral sensibilities. One was *sati* in which well-to-do, high caste widows joined their deceased husband on the cremation fire. Only upper caste families did this and there is no evidence that it was very extensive. The British, however, made it a capital offense in the new laws. *Sati* in the classic form disappeared.

It must be pointed out that this took place at the same time that the British and their cultural descendants, the Americans, were enslaving Africans on a grander scale than any the world had seen. Further, one of the procedures they developed for getting rid of unwanted slaves was to dump them shackled overboard on the high seas. The point is that neither the British nor the Americans could be considered highly moral when their vested interests were at stake.

When Euroman could not take direct action, he exhorted against practices that offended him. One was the lotus foot for upper-class Chinese women, the custom of binding the front part of the foot under the back part until the position became permanent. Although the woman could no longer walk without assistance, she was considered extremely beautiful. After a couple of hundred years of condemnation by Euro-missionaries and other Euro-types, the custom was abandoned.

Thus, all in all, the colored woman's position became more like that of Euro-woman, though by no means exactly the same.

Japan, which became a great industrial and financial power in the 20th century, maintained different standards for its women, about which Euro-Americans continued to self-righteously pontificate. Of course, by then Euro-hegemony was long gone in the Far East, and Americans could only criticize this social evil among themselves. The Americans were too busy trying to compete with the Japanese in economic matters to spend much time on social improprieties.

Other influences on the family system of the colored were of less import except among tribals who had no defense. Sooner or later they had to accept what Euroman wanted. They gradually abandoned their larger kinship connections, moving toward the nuclear family. By the end of the 20th century most tribals no longer knew what larger group they had belonged to. Even the great clan system of China faded out, largely through the influence of their Eastern Euro-creed, Marxism.

As to the other social group of significance, that of special interest, again it was the tribals who were changed most. Though there were some changes before the Euro-inundation, these were few by comparison. American Indian groups had curing, military, and ritual societies which an individual could join. Most of these faded on the reservations as their function ceased. In their place, many new special interest groups were introduced by the Euros.

One of the most significant groups was the Christian Church. Though Christians were often thought of as a single great group, in actuality Christianity had been torn by dissension from its beginning. About the only overall unity Christians had was an acceptance of the Bible as the sacred book and a belief that Jesus Christ was their man-god. Of course, the members of each believed they were "the" most righteous sect and that the others should be converted. Thus, after the Spanish were driven out, practically every reservation had at least two mission churches. Whichever church an Indian joined, his life was irrevocably changed, from the clothes he wore to his sex habits. Over time, his prior religious beliefs weakened or disappeared entirely under the assault of the missionaries who were perhaps even more ethnocentric than other Euromen. To make things even more complex, many new cults evolved which promised to get rid of or make the white man easier to live with. An Indian, a South Sea Islander or an African could join one of these groups or an established Christian church, or even in some cases hold onto old beliefs. Thus, by the beginning of the 20th century, one had to ask a tribal which religion he belonged to just as one had to do with a Euro.

Another important new special interest group to which the tribals had to adjust was the tribal council and its parent organization, the bureau of native affairs. In the U.S. this latter organization was called the Bureau of Indian Affairs. Called the B.I.A. by insiders, it vacillated wildly between advocating cultural integrity and continuing to take rights and land away, usually the latter. Among other things, the officials tried to introduce a council with elected officials, majority rule, and what was thought of as the democratic way. The Indians, for the most part, had different systems and found it very difficult to come to adopt this "new freedom." However, there were councils of sorts for most tribes which varied in effectiveness.

Variations of native affair bureaus were instituted in other tribal areas where the tribes had not been completely eliminated or absorbed. These were involved, for the most part, in acculturating the natives to the Euro-way, inhibiting excessive use of mind-altering Euro-products, particularly alcohol, and helping the white neighbors get what remained of native land or raw products.

Still another special interest group of significance which came later in Euroman's day was the conscript army, quite a different affair from the professional soldier society it replaced. Euroman invariably outlawed existing native wars and military organizations and immediately introduced his own. Tribals all across America were sucked into Euromen's wars and invariably lost much as a consequence.

The Oneida Indians of New York State helped the Americans against the British in the War of Independence. After the war they lost most of their land to the Euro-Americans, and the majority had to leave the state. Euroman used members of one tribe to hunt down those from another, as well as fight his white enemies. The native tracker ultimately helped destroy his own way of life. Many Indians served in World War II. Navaho soldiers were used to speak to one another on radio or telephone so German or Japanese interceptors could not understand them. The Navahos had long been pacified by Kit Carson, and when various Indian agents and veterans returned home, they had to face the power companies which were trying to get the minerals from the reservations for a pittance. It seems that little had changed since the purchase of Manhattan Island by the Dutch.

Tribals had to learn to deal with many other new special interest groups such as the police, courts, jails, and schools as they were sucked into the mainstream of the life of their conquerors.

The great civilizations of the Far East, South Asia, and the Near East, already being complex societies, had their own special interest groups; so Euroman posed no particular problems for them in this regard. They did adopt some new groups and abandon old ones, but they had already been doing this for thousands of years.

12

The Word of God

I flipped a buffalo nickel over and over, marking on a scrap of paper which way it came down. My tally was 18 buffalo, 17 Indian when Mary looked up. She illuminated, "Hi Pete, what's up? You're playing with your coins again. Your mind still on economic matters?"

I smiled. "Actually not. Although their primary function was to facilitate small exchanges, like other cultural things, coins also had other functions."

"I suppose that would be expected of anything that was used a lot.?"

"Usually yes. Through all of my lifetime and before, I'm sure, the one thing you could expect to find on a Euro was money. Those who didn't carry money were generally the very rich or the very poor. Others carried money for the rich and the poor didn't have any."

"And for what other purposes could coins be used?

I replied, in flipping the nickel again. "When we found it difficult to choose between alternatives, we would flip a coin. Heads we do, tails we don't, we used to say." The nickel came down buffalo.

"We would call that tails since it was the opposite of a head. Almost all coins had a human head on one side. Euros liked to commemorate their famous people in this way. This one was unusual with a Plains Indian, a buffalo hunter. They were the ones the Americans took the western states from while destroying the great buffalo herds."

Mary flashed "Pause" for about 30 seconds, presumably to let me know she was considering my statement. Then she came to life again. "I see. The process of decision making. Amazing how many ways you devised to perform this critical task." The screen froze for a few seconds, then, "Is that what you're about to launch into now, Pete, decision making?"

"No, that was not a big issue in anthropology, though some of the other social sciences were quite interested, particularly those which favored the practical use of knowledge. No, I'm afraid my mind was wandering far afield. I'm sure I told you, Mary, that I tended to go off on tangents. When I was in my teaching period, the students who wanted their information straight frequently would become very frustrated with my presentation. I did get a lot of fun though from wandering off on tangents. And I must say, teaching was as much fun for me as it was a totally dedicated effort to pass on knowledge."

I was not surprised when Mary brought me back. "Okay Pete, but let's finish what we're doing. What were you writing when you were flipping the coin?"

"Oh that. I was just keeping a tally of how often the coin came up heads, how often tails. You know I was dedicated to the empirical way, and the results of tossing a coin was thought by statisticians to be a matter of pure chance. Statistics was a branch of mathematics used to prove or disprove scientific theories. However, it was used for much more."

"So what did you find out from your coin tossing? Was it pure chance?"

"It seemed to be. I came up with one more buffalo, but that fit the predictions of the statisticians. Their claim was that if I'd continued to toss enough times, the results would have been almost exactly half and half."

"So then, are we going to talk about statistics now?"

I couldn't help chuckling to myself. Mary was so like the women I had lived with during my couples period of life, invariably trying to second-guess me. I said, "No, believe it or not, my mind had wandered in quite another direction. I was actually thinking about supernatural phenomena."

"So that's where we're going today, eh?" Mary sounded satisfied that she had finally nailed it down.

"I think so, if it's okay with you. But first I think I should tell you what it has to do with coin tossing. There was a connection."

"By all means." Pause. "This comes as no surprise though, Pete."

I chuckled, audibly this time, saying, "Caught again. But what the hell, life goes on. You see, Mary, the part of society I entered in my young manhood, the educated middle class, had become quite committed to the scientific method. They invariably looked for natural explanations of events, as with coin tossing. It was believed that everything could be explained under natural laws. "Oh, they did play around a little with ESP, extra sensory perception. As a matter of fact, there were some experiments to see if thought could influence coin tosses, but by and large empiricists stayed away from explanations or experiments which were not strictly tied to naturalism. And I'm afraid I became one of them."

"Hmm. So what about the supernaturalism you seem to want to talk about. You didn't believe in it?"

I laughed. "I'm afraid not, like most of my colleagues in the profession. We were an interesting bunch in that regard. We spent an inordinate amount of time describing the supernatural beliefs and practices of others, particularly the primitives, and yet the majority of us were non-believers."

Mary delayed for a moment. I figured she was digesting this last bit of information, but she quickly rebounded. "Are you saying that the more you learned about these beliefs and practices, the less you believed them."

"I guess that was partially it. Though I think there was also another reason. Most of us were in anthropology because we were already at odds with our own culture. Somehow we just had not been sufficiently brainwashed." I paused, then, "I explained before how humans were conditioned in a process called enculturation. Well, we anthropogs (I always liked this shorter version better. It sounded much less formidable than anthropologist. And I never could see the students of culture as a formidable lot.) just didn't get conditioned as much as other people, and especially about religion. As a result, many gave up their ancestral religion."

"That happened to you, Pete?"

"I'm afraid it did. I fought my parents tooth and nail about the ancestral faith, Catholicism, and as soon as I left home at the age of 19, I quit totally. Further, I'm sure my attraction to anthropology some six to seven years later was largely due to the fact that it dealt almost exclusively with the ways of others, including their religion." I paused, then came out with the clincher. "As a matter of fact, I think rejection of my own way has been a main theme in my life, along with an acceptance of the way of others, rather then the usual attitude of most people, staying with their own rather than changing to that of others."

"So, Pete the iconoclast, eh? Isn't that what you call a person who goes against the beliefs of his own people?"

"Yes, sometimes. And though I think there's something to that, I hope it's more. I hope that after a life of rich experiences in different cultures, I have come to a stage when I'm as culture free as it is possible to get. That is, now that I can see the negatives and positives in all kinds of cultures, I no longer feel bound to choose my own or any other." At that point I chuckled and for Mary's benefit, I added, "Especially now that none of them are destined to continue in the future."

All of a sudden I felt a loss of energy and was pleased when "Pause" came on the screen. I stayed within that mode of self-analysis until Mary brought me back again.

"It helps, Pete, to know the mental set of one's informant, though now I think we should go on. I'm ready for your views of the supernatural, but I'm still puzzled at the connection with coin tossing. Could you very briefly straighten that out for me?"

I felt relief as I turned back to a discussion of culture again.

"Oh that's not so complex, Mary. It's one of those little gems that came out of the study of primitives. They, like most other people before the age of empiricism, consistently mixed naturalism with supernaturalism. If, for example, there was an exceptionally high tide, they would at times explain it as a consequence of a storm or earthquake, and at other times as being caused by the actions of a god such as Neptune. If someone became sick, it might be thought of as a consequence of eating something that was poisonous, or the consequence of someone's evil eye. And the same applied to material objects such as gambling pieces. Men in the age of empiricism believed that coins could only act according to natural law. Many primitive people, however, thought their gambling pieces, painted sticks or pebbles usually, were subject to influence by spirits. That's what I was thinking as I tossed the coins."

"Ah yes, another of your convoluted mental wanderings, eh Peter Hermann.?"

"I'm afraid so, Mary. As I told you, it's one of my principal weaknesses. But anyway, if you let me go long enough, I usually get to the point. And it's even more likely when I have you to keep me on track. And so now we've come to the meaty part, as they used to say in the carnivorous Euro-American world, the supernatural."

Mary came on quickly then as if she had been saving the question. "Okay Pete., But first tell me clearly what you anthropogs, if I may call you that, meant by the supernatural."

"Okay. You probably don't know that the Euros had divided up the supernatural world into two parts. They called the primary belief systems of civilizations religions, while secondary beliefs, which were frequently remnants of pre-Christian supernaturalism and most of the beliefs of tribals were called superstition. Religion was considered to be true, superstition mere figments of the imagination. In that way the Euros satisfied their ethnocentric tendencies. The problem was that anthropogs mostly studied the belief systems of the pre-urban peoples. And they decided, probably at least partly out of self-interest, that one couldn't logically separate the beliefs of different cultures in that way unless one was an ethnocentric believer. True Christians had no trouble with this division of ideas.

So the anthropogs used the term "supernatural" to mean any beliefs that couldn't be accounted for by natural explanations. It ended up including not only the beliefs of the world's religions, but ideas about the extra-normal which existed in different cultures, the evil eye, voodoo (vodun), fear of a black cat crossing one's path, and throwing salt over one's shoulder after sneezing to avert misfortune."

"All right, so where do we begin?"

"We begin again in the world of the contemporary primitives. As you might guess, evidence of such ideas and the practices which resulted from them were practically nonexistent in the domain of archeology. I believe the earliest evidence cited by anthropogs that mankind had supernatural beliefs was of something that happened no more than 100,000 years ago, about a minute in the five million years of the *sapiens* clock. And the evidence was that man buried his dead. The usual interpretation was that the Neanderthals of that time believed in a life after death because they laid out their corpses in a ritualistic manner. I suppose this was based on the idea that since nonhumans do not bury their dead and so far as we can ascertain do not believe in an afterlife, a creature that did bury its dead must have believed in a life after death. Of course we do not really know that other animals have no such belief, but there is no evidence that it does exist. But still there might have been other reasons for humans to have disposed of the dead in this way. The Marxists, who officially denied the existence of an afterlife, did ceremoniously embalm their famous dead in order to put them on public display, presumably to reinforce the secular creed in the minds of the masses. Also some primitives did not bury their dead. The North Alaskan Eskimo were reported to have left the corpses out on the tundra for the wolves.

"In any event, archeological data is pretty flimsy stuff as compared to the great amount of information on burials that we have garnered from studies of living primitives. So anthropology fell back on the same belief that if contemporary primitives followed a custom, it could be assumed that ancient primitives also did. And there was much information gathered as to what living man believed and did about the supernatural.. Basically, pre-urban man believed in a life after death, but also one far beyond memory in the past, usually to when the world was created. The Australian primitives had a poetic term for this mythical past, "dreamtime." Also pre-urban man believed in worlds beyond the one he inhabited, in the sky, under the earth, and everywhere his imagination could take him. *Sapiens* was the great time and space extender.

"Two special worlds existed: one a reward for the righteous, heaven; and the other a punishment for the wicked, hell.

"Beyond this, pre-urban man had a great variety of beliefs and practices which helped him deal with the problems of this world, such as how to cure illness, how to improve his success in the hunt, how to bring rain, and how to deal with a cantankerous neighbor. Anthropogs called this mana, from the word used by people in the South Seas. Also, primitive man had a densely populated spirit world, the beings of which could either help or harm him. He tried to keep the bad spirits away and get the good spirits to help him. Perhaps the most widespread kind of spirits were those of the natural, nonhuman world. The animals and forces of nature frequently had spirits whose help it was worthwhile to seek.

"Envisioning spirit forces in the birds, fish, and mammals did not prevent primitive man, the hunter, from killing and eating most of these creatures. But even if their flesh was consumed, *sapiens* treated them as he would other spirits, which included allowing them to share his after-world. They were not treated as nothing more than meat.

"Another widespread type of spirit was the ghost of the dead, which was generally harmful.

"There were variations on these themes in early civilizations until a couple of major changes took place. One occurred in India when a religious seeker, Gautama, came to the conclusion that though there clearly was supernatural power, it was useless for weak-minded *sapiens* to try to understand it. He should be content to learn how to eliminate self and merge into the great supernatural nothingness of nirvana before death. Moreover, Gautama provided a formula for doing this, Buddhism, which was to become the main religion of the Orient. Along with Hinduism and Jainism, Buddhism incorporated a belief in reincarnation for all creatures. As a consequence, it was inculcated that nonhuman life was not to be taken at will for food or other reasons. And like the primitives, this did not prevent most Buddhists from eating meat, but because of the concept of reincarnation, nonhuman life was treated differently from that in the Judeo-Christian-Islamic world.

"The other event occurred in the Middle East at about the same time that

Buddhism came into being. This was to have great repercussions on the culture of Euroman, the rise of monotheistic Judaism and its descendant religions, Christianity and Islam. The Jews and Christians believed in one god, basically anthropomorphic, who was exclusively concerned with mankind. The new religion demoted all the rest of animal life into meat for *sapiens* or pests to be eliminated. It banished them from the afterworld and denied them a soul, a supernatural entity claimed to exist in a human being. Christianity rejected almost all the differing supernatural beliefs of others and proclaimed that the only way men would be saved was if they accepted the beliefs and practices of Christians. The Jews had been just as convinced of their views, but weren't concerned with sharing their good fortune with others. The Christians then set out to convert all those they could, and they were quite successful in Europe.

"The other great religion that sought converts appeared after Christianity, Islam. And although the prophet, Mohammed, claimed he'd received his message directly from Allah (God), a usual claim of prophets, the resulting concepts seem to have been drawn primarily from Judaism and Christianity. There was really nothing new in the creed except that Mohammed was the final prophet. Islam shared with Christianity its aggressiveness in spreading the word, and managed to convert hundreds of millions of people before the spread of Euroculture. In fact, 1492, the year the first Euroman officially reached the New World, was one year after the defeat of the followers of Mohammed in Spain which ended the great period of Islamic expansion."

And so when Euroman went forth on the seas of the world, he carried along the creed of Jesus Christ. Although the majority of the first arrivals on foreign shores were soldiers and sailors, from the very beginning there were also clerics or what we sometimes referred to as "men of the cloth."

As soon as any natives were pacified or otherwise intimidated, the first missionaries got busy with their primary task, conversion. Christianity had a special ritual, baptism, by which nonbelievers could be turned into the chosen, at least nominally, by having water poured or sprinkled on them and receiving a new Christian name. It was truly amazing how many people the missionaries managed to baptize in a very short time. The whole of Spanish and Portuguese America, except for a few isolated tribesmen, was made Catholic within a hundred years.

Of course the process was not just a matter of the natives seeing the light. The Spanish and Portuguese came in as conquerors imbued with deep ethnocentrism. And in no area of belief were they more ethnocentric than in religion. Since the natives had extensive religious beliefs and practices of their own, this presented a problem due to the exclusivity of Christianity. Christians had no patience with others' beliefs, and thus saw abominations everywhere. In particular, Christians got very excited about what they called "graven images." In my current dictionary this is defined as "an idol or fetish carved in wood or stone." Both the Jews and the Muslims had refused to allow any portrayal of spirit beings in their temples, especially of their god. The Christians loosened up a little in this regard so that Euroman carried along images of Jesus, Mary his mother, and many saints. As soon as Euroman got in control, he proceeded to destroy the local images. The greatest act of destruction was probably that of the Aztec capital, Tenochtitlan. The Aztecs were a very religious people in their own right and much of their energy and artistry had gone into the making of statues of their gods as well as temples and pyramids. But they also had a deep belief in human sacrifice. Their wars were largely efforts to capture victims for this purpose. And although sacrifice had been known in the Old Testament of the Christians, it had been foresworn in the New Testament. Thus, everything about Aztec religion was an abomination to the Spanish conquerors. And they did what conquerors have usually done, they destroyed what offended them. The Spanish leader, Hernan Cortes, forced the newly conquered Aztecs to go through the great city, Tenochtitlan, and tear down every statue, temple, and pyramid. Afterwards, he did another thing conquerors have frequently done, he had the losers of the war, by then baptized Christians, build Christian churches where the great temples and pyramids had been.

Euro-apologists have justified the destruction by the conquerors on the grounds that the Aztec rituals of human sacrifice were barbaric. Of course, at this same time the Spanish were conducting the Grand Inquisition under church auspices, in the process of which thousands were tortured, garrotted and burned at the stake. And the Aztec population went from 25 to one million in one hundred years under Spanish rule. Moreover, the Crusades were just over, during which thousands, Christians and non-Christians had been killed in the name of Jesus. It is very difficult to see the Aztecs as more barbaric than the Christians.

The central plaza of Mexico City, the Zocalo, was the site of the primary cathedral in the country, built atop a pile of rubble from an Aztec temple. In the 20th century archeologists were continuing to excavate the rubble below the cathedral, searching for artifacts. The same happened in Cuzco, Peru where Christian churches were built atop the foundations of Inca temples.

Other Euromen committed similar acts of destruction elsewhere, usually among pre-urban peoples. American missionaries as part of their efforts to destroy native Hawaiian culture, promoted a split in the royal family, convincing one faction to hurl down the statues of the old gods. This act spelled the end of the Hawaiian religion, leaving nothing more than quaint tales of volcano and sky gods with which to amuse Euro-American tourists. Practically all native Hawaiians became Christians.

The West Africans were one of the great wood carving peoples of the world, in particular carving ancestral figures in great variety and of high artistic quality. Picasso and other Euro-artists based much of their modern art on these figures. African wood carvings became quite valuable to secular Euros by the middle of the 20th century which resulted in West African governments passing laws forbidding their export.

However, when Christian missionaries moved in from the coast in the 18th and 19th centuries, they attacked the native art tradition.

When I did fieldwork in Nigeria in the '70s, few "pagan" villages were left. And wherever the last holdouts accepted the faith of Christ, the new believers were admonished to bring in their wood carvings to be burned. I never saw a traditional native piece anywhere except in museums.

Collectors reaped considerable benefit from the reaction of natives to missionary condemnation of their "graven images." Primitive art, which in the middle of the 20th century became very valuable, could during the heyday of missionization be bought quite cheaply. This was the period of museum building in the Euro-world and great collections of wood and stone carvings were stashed away in dusty basement storerooms.

The wood figures of West Africa came under a double whammy. While they were being destroyed from the south as Christian missionization moved inland, they were being destroyed from the north by men converted to Islam, which also was under the rule banning the graven images, this idea coming from the Old Testament.

Besides the wood figures of West Africa, Islamic conversion brought about the destruction of spirit and deity depictions wherever Muslims gained control. Extensive destruction of Hindu temples and images took place in Hindu India following the Muslim conquest.

There was much less destruction in the old civilizations of the Far East, since Euroman was there precariously, even in his heyday. It was quite a different matter risking an uprising by sizeable numbers of people, as in China, just to destroy some religious figures as contrasted to the low risk of dealing with pre-urban peoples heavy-handedly as in Hawaii.

Once established as conquerors, while conversion was going on at full speed, the missionaries turned their attention to making the converts build churches and other buildings that would be of use in the new religious settlements. The

Spanish were especially good at having great cathedrals built in the style of their home country. But as they extended their reach beyond the centers of power, mainly in Mexico and Peru, the Spanish encountered more sparsely populated regions. There they scaled down their churches and settlements. In the southwestern United States they evolved a self-contained community called a mission in which the workers were native Indians and the managers and religious teachers were Catholic priests and brothers. In California the organizers got Indians from many tribes and wide areas to come to the missions. Since there was no common language, the inter-mixed natives had to learn Spanish. They were also taught how to build European-style buildings, European farming techniques, how to make wine and cure olives, and the new religion. Native practices were discouraged, and in a very short time, all semblance of tribal culture was gone. The Indians lost their own languages and became known by the mission to which they had been assigned: Diegueno, Gabrieleno, Luiseno.

Many bands of Australian aborigines were brought together onto Missions by English missionaries. In fact, the aborigines called the third era of their history "mission time." The one following, which continued until Takeover, was called "government time."

Missionaries were always very concerned with language, their own and that of the people they were trying to convert. This made sense since their message was almost entirely verbal. The basic justification for their existence was "the word," the proclamations of their holy book, the Bible. Their primary method of getting this across was through exhortation, proclaiming the truths loudly or emotionally or both. All rituals depended mainly on verbal formulas. The newborn and newly converted were brought into the faith, errant people confessed their sins and then recited prayers for exoneration, and the dead and dying were sent off into the next world, all mainly with words. By contrast, there were professions, in the applied sciences, for instance, that could depend largely on demonstrating their benefits. If the church had to do that, it would have faced an impossible task. The faithful had to be so on the basis of belief in words.

Indeed, of all specializations in the era of Euroman, none depended more on words than the church. It had to solve the language problem to survive and prosper. Their leaders recognized this early on.

By the time Euroman had gotten involved in overrunning the world, the sacred language of the church had already changed two or three times and the current one, Latin, was already in decline. So when the Euro-missionaries were faced with the need to transmit their message, they did it either by teaching the "others" their own national tongue or they learned the language of the natives. Again the Spaniards took the lead. They taught all their converts Spanish so successfully that except for the survivors of the old Inca Empire in South America and a few isolated tribesmen, Spanish became the exclusive language of the Indians of Latin America.

The tribal groups that maintained their own languages lived mainly in the interior of Brazil and Venezuela. In the 20th century missionaries worked to learn their languages as well as invent writing systems for them. Prayer books and the Bible were the earliest printed books.

Missionaries had done the same elsewhere among the pre-urban peoples, particularly in North America, Africa, and the Pacific Islands. In North America the chief persons on the reservations were the trader, missionary, Indian agent, teacher, and nurse, more or less in that order in terms of importance and sequence of arrival.

Missionaries working in the great civilizations of Asia also became proficient at speaking and writing the local languages, though since these had been literate longer than the cultures of Euroman, there was no need to invent writing systems. However, it certainly must have been worthwhile to exhort the new converts from the pulpit in Urdu or Mandarin.

One of the large groups involved was the China Inland Mission whose missionaries toiled to capture souls while the Euronavies were blasting the cities to force trade compliance in the period of gunboat diplomacy. The Euro-presence, including the China Inland Mission, was totally ejected by the new Communist government in the late '40s.

Missionaries became quite able linguists as a consequence of their vested interest. My first textbook in descriptive linguistics was by such a pair, Pike and Nida. An important center for missionary studies was the Oklahoma Institute of Linguistics. In the '60s I knew American missionaries in Laos who were doing well learning the local tribal languages when practically no other Americans could even speak Lao which had been a literary language for 1,000 years.

Missionaries also brought schools and hospitals to the colored man's world. Primary schools were a natural since it was long known that children can be brainwashed more easily than can adults. What better way to develop converts than Bible study and other religious training, along with the basics of the three R's and European moral values? At the very least, the children had to learn the new writing if they were to use the prayer books or Bible.

In the urban centers of the colored man's world the Christians established universities which disseminated a higher level of knowledge of the Euro-way, along with training in the new Euro-sciences. The American Universities of Cairo and Beirut were well-known for doing this. There also were similar universities in China which were taken over by the new secularized Chinese government in the mid-20th century. The end of the Euro-colonial era brought about a nationalizing of most such Christian enclaves.

One of the oldest functions of the supernatural was to cure the ill. The earliest specialist identified by anthropologists was the shaman who was principally a curer with supernatural power. Curing also remained a big issue in the religions of urban cultures, but in none more than Christianity. Christ was a curer of the afflicted according to the holy book. He lay on his hands and even the dead rose to life.

However, when the Euro-science of medicine developed, it was a secular effort, as were most of the other sciences. Naturalistic explanations did not fit well with the invocation of unseen powers. So the tradition of curing developed in two ways. One was through the totally secular institution which operated for profit or was paid for by the city or state; the university or county hospital became the primary secular type. The other kind of hospital was operated by religious denominations, primarily Christian. In all instances, the medical staff relied exclusively on naturalistic reasoning. In the religion backed hospital, however, there was also much prayer and other invocations of the supernatural.

It was probably inevitable then that the religion-backed hospital would be exported to the cities of the colored man, filling a scientific vacuum, so to speak. Thus, right up to Takeover, some of the best hospitals in foreign countries were Christian-founded institutions, Catholic, Presbyterian, 7th Day Adventist, and other denominations. These were generally available only to the well-to-do in cities, since their fees were high and most of the general populace, especially the rurals, were not familiar with them.

The Christian churches were very concerned with morality, the rightness or wrongness of behavior. However, very few of them worried about the immorality of the dispossession or domination of native populations though a few early missionaries did feel that the methods that had been used were too harsh. But once the natives were dispossessed and powerless, church leaders tended to take their side on civil issues. This has been a characteristic of Euroman generally, but particularly Americans. They have shown no mercy in their efforts to pacify or eliminate native species of animals or people. But once this was accomplished, and the wolves or bears or Indians were either eliminated or pacified, the Euros became very concerned with saving the last of the endangered species and tribal remnants. Thus in the days when Indians were still a threat, Euro-militia were turned loose to destroy their villages, inhabited by mostly women and children, and the last male holdouts were designated as renegades and hunted down. And once the spirit of the Indians was broken and the "Redskins" established as the lowest economic class of the new society, it became quite fashionable to make documentary films of their plight and to write sympathetic accounts for publication. The same was true of the great beasts wolves, bears, buffalo, and elk. Once they were brought to the edge of extinction by guns, traps, and poison, it became "the thing to do" to save a few of the survivors. The Indians got reserves, the big game animals, preserves.

Further, while Euroman was in the business of taking over the cultures of others, little time was wasted worrying about their welfare. Only later, after they were no longer competitors, could he become considerate.

And though perhaps from an insider's point of view the greatest immorality of Euroman was taking over the land of the weaker others, he was little concerned with this. Rather, it was the human body and its functions that bothered the cleric, once the "graven image" problem was settled. Then, usually after the natives had been converted and were going to church, he got busy trying to stop immoral behavior, by his standards, of course. The uncovered breasts of females particularly bothered him and he worked diligently until he got them covered, mostly before the middle of the 20th century. A Euro-invention, the brassiere, a female undergarment designed to cover the breasts, came in handy. In many parts of the tropical world women who formerly had worn nothing over their breasts, donned the bra as an outer garment.

Many years ago in the New Yorker magazine there was a cartoon showing a bare-breasted, dark-skinned woman running up to a group of women, a brassiere in hand, saying, "Quick, take off your brassieres. Here comes the National Geographic photographer."

Other parts of the female body had to be covered up also according to the missionary. Arms and legs were temptations and women were required to wear long-sleeved, long-hemmed dresses. Dresses such as the muu-muu took care of the midriff as well. Even the head could cause problems, particularly the hair. Since women could adorn or style it to make themselves sexually attractive, a head covering had to be worn in the Catholic church. Rouge and powder and other facial coloration also was discouraged.

The sister religion, Islam, also faced the problem of females wearing immoral attire. But the Muslims went even further, keeping women out of the place of worship, the mosque, and making sure they were well covered when outside the home. One of my Hindu friends called this outfit, "a walking tent." Even the face was covered in classic purdah.

Other kinds of sexual immorality were attacked by Christian missionaries, including fornication and adultery. All "unnatural" relations such as polygamy, wife swapping, and homosexuality were condemned as immoral.

However, the world of Euroman changed considerably in the 500 years of his influence on the colored man. By the end of the period of Euro-dominance, the notion of love and romance had gained ascendancy, along with a decline in religiosity. Thus, at the end of this era the colored man found Euroman approving such open expressions of affection as kissing and hugging between the sexes, as well as much body exposure and even acts of sexuality, especially in the cinema. Even the clerics became more liberal. Some ministers condoned homosexuality while others approved of contraceptives for unmarried teenagers. Kissing in public became commonplace. This was a new moral reality for the colored man.

The effect of the Christian missionary effort varied considerably in different countries. In the long-established civilizations with major religions of their own, the missionary had to be content to convert a small minority. This was true of China, Japan, Korea, most of the Islamic world, and India. The missionary learned early that such people did not switch religions easily. So he concentrated on the tribals or lower castes. Without the support of strong, central authorities, and seeing their cultures degenerate on almost all other fronts, the tribal peoples were usually willing to accept the supernatural creed of the powerful outsiders. Thus, most of the American Indians, Australian aboriginals, Africans south of the Sahara, and people of the Pacific Islands became Christians.

One group that became most dedicated were the descendants of African slaves, the American blacks. Torn from their tribal homeland, they were shipped wherever the slaveholders wanted them irrespective of their tribal background. Thus, in addition to having no common language, most of their customs were obliterated by their Euro-masters. Those from different tribes frequently did not understand the customs of others. Thus the slaves came to speak the language and follow the customs of their masters.

In the United States they learned to speak a special dialect of English. What they ate, later called "soul food" to give it a dignity it originally lacked, was made up of cooked grains, especially from corn, leafy vegetables, pork fat and the organ meats of animals, the least desirable food of their masters.

The slaves also got their masters' religion, Christianity. All Africans had well-developed religions in their homelands, mainly forms of ancestor worship, but with much else. However, in the slave quarters the tribes were so mixed that what one believed was unlikely even to be familiar to another. Further, the Euro-masters discouraged or condemned practically all native practices. The upshot was that apart from the retention of some magic and folklore, the slaves lost their African religious heritage.

In the meantime, they were exposed to the religion of the new faith of Jesus in which, among other things, equality under god was espoused. The contradiction inherent in the fact that white Christians were holding black Christians in bondage was only one of many in American history. In any event, the blacks adopted the religion of the Israelite with a vengeance, becoming staunch believers, probably more devout than their masters. They added much of their own, including musical styles and a kind of emotionalism not then found in Euro-churches. For the most part they continued to be segregated from their masters in the church and elsewhere. However, when peoples' rights became an important issue in the middle of the 20th century, much of the black leadership came from the church. The minister-leaders called for freedom in the name of Jesus.

Human groups usually have resisted encroachment by others, at first through physical means. The colored men of the world were no different. There was much armed resistance from the time Cortes attacked the Aztecs until the United States sent troops into the Philippines. Throughout the world, in nation after nation, colored men learned that armed resistance did not pay off. The Euros invariably suppressed the revolts, and this usually was followed by reprisals.

The conquered then tended to accept their subordinate status apathetically, deriving consolation from some of the new products introduced by Euroman such as alcohol, opium, tea, coffee, and sugar. Since their whole way of life changed, they must have appreciated the mind-numbing quality of alcohol. But some fell back on the supernatural, invoking spirit forces for help.

An action was started in many parts of the world, the aim of which was to restore the old ways. Anthropogs called it a revitalization movement. Typically, after the native group had been trounced by Euro-forces, a prophet would appear. He would claim to have received messages from some spirit being who told him what the people had to do to restore the old order. This would include getting rid of the Euros or making them easy to tolerate. The message usually included injunctions to be morally upright, as well as to perform some special rituals. Generally, one was supposed to quit drinking and stop being sexually promiscuous.

There were dozens of such nativistic revivals in the wake of Euroman's takeover in the South Pacific, Africa, and both Americas. The North American Indians came up with several, the Ghost Dance, Handsome Lake Cult, Bole Maru Cult, and the Peyote Cult.

Euroman did not take kindly to these movements, which almost always appeared after the natives had been pacified. In the case of the North American Indians, the revitalization movements spread after they had been confined to the reservations. Just when the Indian problem seemed to be settled, the religious movement would appear. Specific practices which bothered the whites were dancing for long periods or taking a hallucinogenic drug, peyote, as the gift of Jesus/Thunderbird. In other parts of the world there were similar practices in new cults.

Since Euroman was always a man of action, he ordinarily took steps to suppress such movements, usually outlawing them and incarcerating the leaders. Sometimes he took military action. The Battle of Wounded Knee was actually a massacre of Indians who were performing a Ghost Dance ritual.

One movement survived despite all opposition, the Peyote Cult, which was officially registered with the U.S. government as the Native American Church. This was a creed by which the Indians could live with the white man as brothers. The central act of the ritual was the taking of the peyote button which was claimed to be a special gift to Indians from god. The drug in peyote, mescaline, was reputedly not harmful, though it did have some unpleasant side effects while giving the worshipper a high. Probably the reason the cult was not outlawed by the Euros was that it espoused brotherly love and even accepted Jesus as a legitimate spirit.

When they first heard of these movements, most Euros thought they were weird and perhaps dangerous. Few recognized that their own religion, Christianity, got started in a similar way. The Jews had been suffering at the hands of various higher cultures the Egyptians, Babylonians, Assyrians, and finally the Romans for hundreds of years when a prophet appeared with a message on how to save them. The Romans found out that the Nazarene was inciting the populace to follow the new ritual procedures. The Romans did the usual: they got rid of the prophet. Jesus was executed. Unfortunately for the Romans, the Christians were already on a conversion roll, and despite Roman efforts to eliminate them, they succeeded in undercutting the empire and establishing themselves in the center of Roman power. The moral was that although most such movements were suppressed, those that did survive often became major religions.

Of course, the Christians claimed that their prophet truly knew the word of god. But Mohammed, another prophet, was just as certain. So it should come as no surprise that the followers of other revitalization movements believed their prophets as well.

13

Other Customs

I rummaged around in the little alcove, studying the dusty odds and ends. Most of them seemed to be backdrops, fragments of scenes. I tipped the various painted panels forward and continued. All of a sudden I saw one which was slate grey, and under the dust I could vaguely make out white marks. I became interested, not knowing exactly why. Anyway, I quickly tipped the rest of the panels forward and freed the slate one. As I slid it out, I knew it was what I had thought. I tried to dust it off with the only things I had available, my hand and sleeve. As the dust brushed away, I saw that the white chalk marks were in some kind of formation, little x's lined up with larger ones in the middle and ends, identified by sex [♂ or ♀]. Most of the little ones were female [♀]. Arrows showed the x's which way to turn, advance, or go back. Wow, I thought, a dance score by someone who knew the bio-symbols for gender.

I pulled the chalkboard out and saw that it was a self-contained unit with folded legs attached. I set it in position and dusted it.

Mary's screen lit up. She actually laughed, the first time I remembered hearing the sound. I looked up and on the screen was written, "HA, HA, HA."

I have a feeling that I reddened though I felt this wasn't appropriate behavior for an almost-centenarian. Blushing was for inexperienced youth.

She said, "What in heaven's name are you up to this time, grubbing around in that dusty alcove?"

"Oh, you are still in for a few surprises, Mary." I wanted to say "old girl" but couldn't bring myself to it, both because she was one of "them," even if only an electronic extension, and I still just could not think of a computer as having a gender.

"Ha, ha, ha," again, in lower case. "I'm sure I am, Pete. And that's okay. After all, if I had known all the hijinks of *sapiens*, we would hardly be doing what we are, right?"

"Right, Mary." I noticed that I used her name a lot more than I had in the beginning, as she did mine. "So I guess you want to know why I am getting out a chalkboard? But before I tell you, let me ask you something. How did all these panels and the chalkboard get here?"

Mary hesitated, then, "I'm afraid I don't know myself, Pete, but I can check it out in the memory section. You know the Atierrans set up a massive information retrieval and storage system, including what existed before reconstruction. Wait just a minute." Her computer screen flipped to "Processing" and held steady.

I made myself busy looking for the one other vital piece for my board, some chalk. I was rewarded by finding a piece about two inches long. That would be enough for what I was going to do. When I straightened up and dusted off, Mary's screen began to write. "Previous structure was live entertainment center, what was called theatre in English. Panels remaining were backdrops for productions."

So that was it. The computer center had been built over an old theatre, probably summer stock or amateur. That also explained the dance diagrams. The last production had been a musical and the dance director had been explaining a number on the chalkboard. How appropriate for me.

Mary came on in her interview mode. "That sound reasonable, Pete, a theatre?"

"Yes, and not only that, but it is most appropriate for the direction I want us to go in the next interview, as well as having warmed the cockles of this old professor's heart."

"You sound very pleased, Pete. Does a chalkboard and piece of chalk do that much for you?"

"It sure does. You know, Mary, I was a professor for many years and these were my basic props. Ostensibly the chalkboard was a device for allowing students in large classrooms to see the magic symbols, words, numbers or miscellaneous cryptographs from anywhere in a large room. And the more bodies colleges could pack into a classroom, the more money they made.

In actuality the chalkboard became a prop for many professors who would write gibberish on it just to have something dramatic to do while they were lecturing. I knew a professor of folklore once who would start by drawing an arrow and throughout his lecture continue to add to the tail of the arrow and change direction so that in the end the board would be covered with an arrow that angled off in all directions and explained nothing. The dedicated student would put the original version of the arrow in his/her notebook but soon give up, seeing no meaning in it. The lecture, I might say, was usually stimulating, if not always comprehensible.

"Was that your style, Pete?"

"I suppose I did do some of the meaningless stuff, though with me it was almost always words. I was a totally dedicated wordsmith." I paused, remembering. "My worst fault was leaving a word dangle in nowheresville. I would get so anxious to get on with my point in the lecture that I wouldn't take time to finish the sentence. Anyway, that's not what I'm going to do with you, Mary. I'm going to present an idea that became popular in anthropology in the form of a diagram. And we will follow up the idea in our talk today, okay?"

"Sure, whatever will make the old professor happy." Mary was getting playful with me which was okay. Despite having become an old codger, I had never lost my pleasure in playing around. I stepped back and drew the figure.

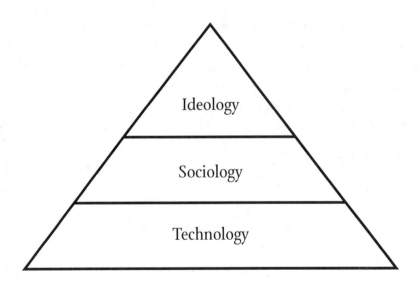

"So, a triangle with three levels or tiers. Is it a pyramid, you know like the ones built by the Egyptians and ancient Mexicans?"

"It is in the shape of a pyramid. but so far as I know that is coincidental. The figure is supposed to represent the major areas of any culture, the most vital part at the bottom, the lesser at the top. It has quite a story, being the product of one of our best known curmudgeons of the generation before mine."

"Ah, theory again."

"'Fraid so, Mary. But you can relax, there's nothing very complicated about it, even if it does go counter to what many people think. It's always been some help to me in understanding cultural systems. I hope it will help you. And if it doesn't, it won't use up much time, after which we'll go on to the topic of the day anyway. Okay?"

"Sure Pete. So why don't you explain the meaning of the triangle?"

I stood back to view the figure, but came forward to tap the board whenever I was making a significant point. "First though, I think I should tell you about the curmudgeon to whom the diagram is attributed. He loved to make waves, and was known as a neo-evolutionist because he came along just at the waning of the period of salvage ethnology during which ethnographers were busy trying to record the remaining information of the American Indian cultures. Evolutionism, and most other theorizing, were not permitted by Papa Boas, the grandfather of American anthropology.

"Anyway, old Leslie White, the diagram's father, didn't like all that business about learning every detail about every little culture that ever existed. He decided we needed hotshot theorists too and decided he was one of them. Moreover he decided that the old evolutionists had been right more or less, and he would restore their reputations. So he worked out the system presented in this diagram. He claimed that cultures changed to increase their efficiency, to better adapt to their environment, and that was how they progressed. Moreover, he maintained that what was most important was a culture's technology, its tool-making capacity. This would be Tier One on the diagram. Second most important would be the social organization of the group or its sociology. That would be Tier Two. And least important would be the idea system or the ideology. That would be Tier Three. Professor White spoke most often about religion as Tier Three, though I think of it as all the rest of a culture, including its customs for play, entertainment, and the arts."

"I see. And you say he claimed that cultures changed progressively primarily on level one and least on level three?"

"Yes, technology would be most likely to be progressive, he would say. That is, a culture that depended on animal power would be more progressive than one which depended on human power. And one which depended on machines would be more progressive than one which depended on animals."

"And the same would be true of social organization? Cultures that had elaborate social systems would be more progressive than simpler ones?"

"Well, more or less, although it wasn't presented all that clearly. Leslie was mostly interested in technology. But the general idea was that a social system automatically became more complex as the technology became more efficient. And in general, I think that was true. An industrial city of millions was certainly more complex socially than a primitive hunting and gathering band of 20 to 30 people."

"Okay, no quarrel. And that takes care of Tiers I and II. But what about Tier III what you say Leslie called Ideology? I can guess that you are going to say that your professor claimed it was even less significant?"

"You guessed it, Mary. Only apart from the supernatural, Leslie was so un-interested in this level, he hardly bothered about it. Religion was a case apart to him, I presume partially because he was an avowed nonbeliever. He argued that as far as progress was concerned, it didn't matter what a people's religion was. And in a sense he was right. You know we *sapiens* ended up with three great industrial powers in the 20th century and their religious inclinations were to-tally different. Euroman, that is the Western version, was basically Christian with a lot of nonbelievers while Eastern Euroman was driven by a creed of nonbelief, or Marxism, with a minority of oldsters hanging on to Orthodox Christianity. The third great industrial power, Japan and the other Far Eastern-ers, were Buddhist, Confucianist, Shintoist eclectics. In other words, success in the industrial world seemed to have little correlation with belief or nonbelief in the supernatural."

"Interesting Pete. But I thought we had already talked about religion."

"We did. This is a kind of backtrack. But rest assured, I do not intend going over that again. However, you will remember I said that I thought Leslie's Third Tier included many more customs than simply idea systems. And I still do. In fact, although I agree with the old boy about religion being relatively less impor-tant, it certainly did contain a lot of customs that people thought were impor-tant. That is, while there is no doubt that Euroman conquered the world by means of firearms, seagoing ships, and iron armor, he still thought his greatest achievements was his religion, music, and art. And so did others. The Arabs were probably prouder of their calligraphy and poetry, the Hindus of their clas-sical music, and the Chinese of their porcelain and paintings than any of them were proud of their conquests. They thought of these artistic activities as being civilized while conquests were merely necessary because the natives weren't act-ing right. In other words, *sapiens* was prouder of his achievements in Tier III than in Tiers I and II.

"I think I am getting your message, Pete. You are telling me that what men did in technology and social organization was more important for their survival in the real world than what they did in the arts. And yet despite this, *sapiens* valued his arts the highest, and I might add, his religion."

"Yes, that's it, though I would include some other kinds of customs, for instance those for play and entertainment, and some other ways that have hardly been categorized."

"And that's where we are going today into Tier Three?"

"If it's okay, Mary. After that I think we will have covered most of the waterfront, so to speak. We will have pretty much nailed down culture and what Euroman did to it worldwide."

"Okay, let's go. And as usual what about the early, days? You can skip the part about the lack of direct evidence from archeology. I know that what you will be telling me will be mainly about the last surviving primitives."

I chuckled. Mary and I were really getting to understand one another. "As you know, we already did religion or the supernatural so we'll start with a basic we haven't discussed, play. Even the nonhumans did that, horsed around with no apparent goal except to have fun. All the young nonhuman primates played, mostly in wrestling games. The monkey watchers concluded they did this to build up their muscles and reflexes so when they were adults they would be able to either flee from their enemies or put up a good fight for sexual privileges. The primatologists noticed that when these animals grew up, they slowed or stopped most active play. *Sapiens* may have slowed down in old age also, but he continued some games right up until he became decrepit. So game playing came to be a very human kind of behavior. There were a great variety of games for adults up to 1492.

"Another type of custom among *sapiens* which must belong to Tier III seemed to have no primate base, but existed among all peoples we knew of, decoration. It didn't matter how primitive a people was technologically, they invariably tried to make their surroundings more pleasing by adding color and/ or designs. No humans that I learned about were satisfied with things as they came naturally; they tried to improve them. Any undecorated surface would do.

"Since the human body was the most convenient object at hand pre-European man proceeded to improve it through decoration. He painted it, tattooed it, scarred it symmetrically, stretched parts of it, changed the shape of some bones, perforated it all over, and hung decorations on any parts that would so serve. And finally he put decorative clothing on it. Clothing undoubtedly served many functions, including protection from the extremes of climate and to mark one's social status. But in most cultures it also fulfilled an esthetic need. People tried to be handsome or beautiful through the clothes they wore.

"Many objects of most cultures were also subject to decoration, pottery, baskets, animal skins, wood, and stone. Way back when, and this comes from archeology, *sapiens* started to paint natural surfaces, particularly rocks. And later on he began to prepare flat surfaces, or use those of constructions, on which to paint his designs.

"One of the most widespread kinds of construction was the dwelling place, originally to protect *sapiens* from the elements. But such buildings soon became more elaborate, probably most often to be used as an abode for spirits or a place to worship them. These buildings also could be used to satisfy esthetic needs both in form and decoration. Thus the art of architecture came into its own in the urban cultures of the world.

"And though the most important sense organ of *sapiens* was the eye, the ear lent itself to esthetic pleasure in the form of music. Most everywhere in the world men made and played musical instruments. They started with the voice and went on to construct the musical bow, and the hollowed-out log drum. These of course became more elaborate through cultural development that was mainly of four types: the vocal, the stringed, the percussion, and the wind instrument. Gradually, various tonal and rhythmic patterns evolved.

"Another source of pleasure came from taking into the body a wide variety of plant products which raised or lowered the feeling threshold. or gave the consumer visions of other realities, the various drugs. The primary ones during pre-European times were alcohol and tobacco.

"There were other customs which though they might be placed in one of the other categories, could also be considered "odds and ends" or miscellaneous. And somehow this seems proper. No way of classifying is perfect. So some other customs were methods of noting time, mapping, and greeting.

"And when Euroman stepped off the home shores he too had cultural baggage of Tier III."

When he got settled on foreign soil and had the serious business of conquest under control, Euroman pulled out his various games so that he could enjoy the fruits of his labor and also foist them off on the natives.

A few games went in the opposite direction, that is, from the natives to Euroman. But according to the laws of cultural change, most new customs went from the conquerors to the conquered. Two games that Euroman took over from the natives were lacrosse, a North American Indian game, and polo, a game of horsemanship from Southeast Asia. Other games of the natives were hardly considered.

The Anglo-Euros were particularly sports-minded, and spread their games far and wide as they went into the colored man's world. The Anglos were particularly fond of ball games. The basic game of pitch and bat cricket was spread throughout the colonized countries so that in time, some of the greatest teams and players were Africans, Pakistanis, Indians, and Caribbean islanders. And although most adopters of the new games took them pretty much as they were presented, some natives changed cricket to suit their own needs. The people of an island group in the South Pacific, the Trobriands, evolved their own kind of cricket to replace their pattern of warfare which had, of course, been outlawed by the Euros. Cricket had been introduced to them by English missionaries after the natives had been pacified by the English military.

Another ball game that Euroman spread widely was soccer or *futbol*, a rugged game of driving an inflated ball across a field by kicks or body blows. The Spanish spread it through the countries they conquered, including Mexico where an Aztec form of football had already existed. The Aztec game, of course, disappeared when the Spaniards took over. The English spread soccer throughout their colonies so that the great soccer players of the 20th century were from mostly the same countries as the great cricket players.

English soccer evolved into American football which was not spread to other countries, probably because they were already used to soccer. However, two other Euro-American ball games did go to lands of the colored men. One was basketball which involved tossing an inflated ball about the same size as that for soccer into an overhead basket open at the bottom. This game got to be identified with the colored man even before it was exported to other countries. American blacks, the descendants of African slaves, who had many very tall men among them, turned out to be very good at tossing the ball into the overhead basket. Their height alone gave them an advantage. However, the American game of basketball was taken to many other countries that also developed teams that were quite good.

The other American ball game which became very popular in other cultures was baseball, an outgrowth of the English game of cricket. It involved batting a small hard ball as far as possible without being caught, the batter then running a diamond-shaped pattern in stages. And though the Anglo-Americans remained the primary players of baseball, the game did catch on in several other countries, especially in the Caribbean islands. Puerto Rico and Cuba became well known for their baseball teams. But the game also fascinated the Japanese who developed it into a veritable cult, redesigning its social patterns to fit their own needs.

A particular kind of tournament, though invented by the ancient Greeks, was revived by Euroman. This was the International Olympics, originally a series of contests consisting of the major athletic events and dances of the time. And as was characteristic of public life in ancient Greece, it was strictly a male affair. Women were not even allowed as spectators. However, when it was revived by a Frenchman in the 19th century, it was de-sexed, internationalized, politicized, and broadened. The Olympics became the most highly publicized sporting event of the 20th century, attracting the premier athletes of the world, including women.

Euroman, the redoubtable conqueror, was by no means immune to the widespread human tendency of decoration. This, of course, does not mean that he found the decorative efforts of others to his taste, as was typical of his attitude about so many other customs. Wherever Euroman took over, he tended to discourage the practice of body painting, and in most places the local style rapidly disappeared. The use of body paints among tribals was not usually stopped by force as were other customs such as native warfare or local punishments. Instead Euroman inculcated the idea that painting designs on the body was not civilized. Usually, it disappeared when native warfare was abolished and the natives converted to Christianity. Also, of course, the new custom of covering the body made such decoration less significant. What native wanted the gorgeous designs to be covered by a loose cotton dress or baggy pants and shirt? Eventually, body decoration was used only on ceremonial occasions. In general, as the natives became Christians, they abandoned most of their own rituals and thus the body paint.

Even the cultures which survived the Euro-onslaught were affected by this bias. Japanese women, for instance, had used white powder extensively in pre-Meiji times but this declined significantly as Euro-fashions took hold.

But as Euroman effectively discouraged the use of body paint, he and his women introduced decoration of the face. Though men had used powder and a few pigments early in the Euro-era, by its end face paint was used almost exclusively by women. A wide variety of paints were used, mainly on the lips, cheeks, and around the eyes. My wife's ritual, which she called "putting on her face", took at least an hour each morning. After her morning coffee she would station herself in the bathroom to brush, dab, and powder the critical indentations and projections until it looked proper.

As this influence reached most parts of the world, the women adopted face paint, particularly lipstick. However, in some places, skin color was too dark for the paint to show effectively. This was particularly true of Africans, South Asians, and some peoples of the South Pacific. But they simply used more, again mainly on the lips.

Euroman discovered that the natives of the Pacific were using a form of permanent body decoration, tattooing. The word is of Polynesian origin. This was the technique of perforating the skin, rubbing in pigment, then allowing the skin to grow scar tissue over the coloration. Sometimes elaborate designs covered practically the entire body. The people of the central Pacific islands, as well as many East Asians, had relatively light skins so tattooing showed up well. The seafaring Euromen were first attracted to tattooing when they saw the handsome, decorated bodies of the Pacific islanders. So they adopted the idea, one of the few Polynesian customs that did survive. Thus, tattooing became associated with sailors and manliness in the Euro-world. Elsewhere, it served a variety of other functions. In some parts of the Buddhist world tattoos took on a religious connotation. Very devout persons would have their bodies covered with the positions and sayings of the "enlightened one." In Japan tattooing became an identification marker for men of the organized underworld.

Anyway, tattooing spread in the Euro-world, particularly the United States, from sailors to the working class to motorcycle riders, and ultimately to social protestors or romantics who could make "statements" with their tattoos. The center of tattooing shifted to Long Beach, California, where the technique was also sanitized and electrified. The painful ordeal of primitive tattooing was replaced by a relatively painless electric punching and coloring. And though tattooing did survive in the Euro-world, it got a reputation of being appropriate only for the lower classes or social protestors.

The various forms of body deformation practiced by the "others" were generally discouraged by Euroman. The pointy or sugar-loaf head of some American Indians, formed by binding the infant's head, disappeared in the early stages of Euro-influence. The stretched lip of Central Africa and the stretched neck of some tribal Burmese took a little longer to eliminate. The lotus foot of upper-class Chinese females lasted into the 20th century.

Most of the perforations of body parts also went the way of the Dodo. Lip and nose plugs were practically gone by the end of the 20th century. The one place where nose rings survived was India where the well-dressed female continued to wear a small, jewelled ornament in one perforated nostril.

The part of the face that was favored by Euroman/woman for hanging a decorative object was the earlobe. This was a natural since the little tab of flesh served no other particular purpose. Many colored men of the world had been using it to hang decorations when Euroman arrived. A large hole was sometimes made to hold an ear-plug, sometimes 2-3 inches in diameter. The royalty of the Incas of South America, perforated their earlobes and hung weights so heavy that the flesh stretched a foot or more. They were called by the Spaniards "orejones," meaning "big ears." Needless to say, this kind of ornamentation was quickly discouraged by Euroman and rapidly disappeared.

In its place, the Euro-style of using a small perforation or set screw in the earlobe to hold an ornament was introduced, and by the end of the 20th century it was the predominant hanging decoration of females worldwide. Some male groups, particularly homosexuals, took over the earring as a social marker and beauty aid. With the ear ornament, one homosexual could recognize another without going through a lot of preliminaries. Also offbeat groups, particularly females, developed procedures for hanging rings or other ornaments on projecting body parts that were usually covered up, the navel, genitals, and nipples. And then clothing. In addition to its usual functions, it could be decorative. Through the 18th century, Euro-males wore fancy ruffled clothing, powdered wigs, and powder on their faces. A popular addition was a lace handkerchief. Females at that time also dressed up decoratively. However, by the end of the 19th century the Euro-male was becoming more subdued in attire while the female was becoming more flamboyant in fashion.

The Euro-male continued to decorate his head. The high beaverskin hat became *de rigeur* in the 19th century, and a little later, the felt homburg. A major part of the fur trade of North America was trapping beavers for such hats. One of the most pathetic kinds of photo of the late 19th century were those depicting a tribal male, garbed in a breech-clout or makeshift Euro-pants and shirt, but wearing on his head (uncomfortably) a top hat. Frequently he was on a visit to "the great white father" in Washington, trying to get some of his land back or at least trying to keep from losing more.

Anyway, as Euro-influence intensified, the native costumes were replaced by European styles. So while the free North American Indian wore a buckskin breech-clout and leggings, a buckskin shirt, and an over-the-shoulder animal skin robe for warmth, the reservation Indian dressed in tight-fitting denim pants, shirt, and jacket, plus a cowboy hat. His woman traded her wraparound or apron and buttocks cover for a 19th century ankle-length dress or skirt and blouse. The Australian aboriginal male went from wearing nothing to tight pants, shirt, and Aussie hat, his woman going from nothing to a long, ill fitting dress.

Everywhere, the tribal native dressed in some kind of uncomfortable combination of what he could get and what he thought was proper. Needless to say, he generally looked pretty seedy. The Indian got some whiteman's clothes at the trading post as soon as he had enough beaver skins to trade for them.

There was no choice for the little ones who were sent to Indian schools when the reservation system went into effect. They were forced to wear whiteman's clothes as well as speak English and eat with a knife and fork. There were graphic photos of these Indian tykes, garbed in misbegotten combinations and in a state of semi-shock from having been torn away from everything they had known by white militia or cavalry. There were several instances when Indian children were rounded up by federal troops and forcibly carted off to the new Indian schools.

In the 20th century it became fashionable for churches to collect clothing for Indians. Of course, by then they were fully pacified Christians, living on small reservations.

So, for better or worse, almost all the ex-tribals of the world were in Euro-clothes by the end of the 20th century. For men, these were pants, shirt, and jacket where it was cold, and loose dress or skirt and blouse for women. Also, as Eurowomen began to wear different types of pants in the middle of the 20th century, tribal women began to do the same. The most common type of pants was blue jeans, tight fitting, blue cotton pants, used by cowboys first but becoming a unisex, informal costume all over the world by the end of the 20th century.

The Euros developed a formal outfit for males by the 19th century, the suit. It consisted of pants, jacket, sometimes a vest, usually of wool, plus a cotton shirt and tie, a decorative strip of cloth that hung down from the neck over the shirt. This came to be the costume of business and international affairs. However many of the colored men would put on their native attire when they returned home at night. It was inconceivable for anyone, colored or white, to try to conduct business in other clothes. Later on in the 20th century, after the waning of Euro-power, the Africans did on occasion wear a garment that was a tribal version of one of their groups. It was called a dashiki. Arabs from the most conservative countries would sometimes go to business or diplomatic meetings in their long, flowing robes and cotton head coverings. However, many of them had control of so much oil that others, including Euros, would deal with them no matter what they were wearing.

A variant of the Euro-costume was worn by the Chinese for 40-50 years. When they adopted Marxism, they took over the costume of what they conceived as that of the workers of the world, black or dark blue, baggy pants and shirt, with a padded jacket for cold weather and a billed cap of the same material. This costume won no prizes in fashion shows.

The Euro-female became deeply involved in decorative fashion by the end of the 20th century. Elaborate displays, called fashion shows, were started in the Euro-world, but also became popular in cities of many nations where colored women were wearing Euro-styles. The wildest designs and color combinations were promulgated, and these had to change every one or two years. The world of female fashion was largely an economic matter to the organizers, so it was imperative that women change their styles frequently.

In sum, then, the Euro-suit became the international garment of business and diplomacy while the high fashion dress became the public garment of urban females except in the Arab and Indian world and among the early Asian Marxists. Country women continued to wear some semblance of their own costumes or an earlier version of Euro-garments.

There was a distinction in the Euro-world between fine and common arts. In general, body decoration of all sorts was considered a common art, if an art at all. Fine arts were those which had no other function except to satisfy the esthetic need. In practice, the fine arts were created by professional types called artists and they were considered appropriate to display in a fine arts museum. They came into their own in urban cultures, particularly in the Euro-world.

The Euros were very proud of their fine arts. They considered these a prime index for being civilized. The fact that Euros had influenced others so powerfully primarily because they had guns was not something for which they wanted recognition. But the fact that they produced art that others would accept made them proud.

Fine arts were a product of cultures where there was extensive division of labor and where the specialist called an artist could devote all his time to their production. But the tribal peoples used their talents primarily to decorate utilitarian objects. The primitive potter decorated her water jar, the hunter's wife decorated her husband's buckskin garments with porcupine quills, and the wood carver turned his talents to making ancestor figures for worship. Later in the Euro-era some of the tribal productions became popular enough so that they could be made for their own sake, primarily to sell to Euroman. The Navaho blanket and Zuni pottery were made to sell to Whites. These crafts were considered to be between common and fine arts.

The great civilizations of Asia had their own fine arts, most of which continued into the Euro-era though their production waned. The East Asians in particular had brought painting, sculpture, and porcelain to a high artistic level. However, they declined in quality as the Euro-era progressed.

In the meantime Euro-art also was going through transformations. At the beginning of the Euro-era religious paintings and carvings were in style. But as Euroman's faith in Christianity became weaker, a secular, realistic style became more popular, landscapes, portraits, still life, and other worldly subjects. Then in the 20th century a style which was called impressionistic or nonobjective evolved. It depicted the artist 's subjective view of the world, with color and form the main concern. This was a logical outgrowth of the individualism that

176

had permeated the Euro-world by that time.

Many viewers often did not understand the inner vision of the artists, and a common reaction of the public to modern art in the 20th century was, "What does it mean?"

Artists, as a rule, refused to explain their productions, assuming such lack of understanding was the public's loss.

Anyway, this way of painting was exported to the art schools of the rest of the world where it took a place alongside what remained of traditional art styles. It did not become as important as in the Euro-world, but did have an effect.

Another custom of mankind with a strong esthetic component was architecture. Originally, buildings were put up merely for shelter, but even in primitive societies men began to decorate them. The Plains Indians painted pictures on their tipis while the Northwest Coast Indians carved figures on their houses and on poles erected outside.

Urban man took even greater pains with decoration and symbolic construction, in particular for religious purposes. The Aztecs, Incas, and Mayas were famous for their ceremonial architecture as were the Persians, Babylonians, Egyptians, Greeks, and Romans. The same was true in the Asian world, great architecture being incorporated into the religious ritual of Hindus, Buddhists, and Muslims.

When Euroman went forth to the lands of the heathen, he took along the architectural concepts of Christianity. And wherever he conquered an urban culture, particularly in the Americas, he destroyed their places of worship and had the locals build Christian churches in their place. The great cathedral of Mexico City stands upon the ruins of the central Aztec pyramid while the main cathedral of Cuzco, Peru, stands on the ruins of an Inca building.

In places where his conquest was not so total, Euroman simply built churches alongside the local religious buildings. This was generally the case in Asia.

Euroman's churches were less than welcome in the Islamic world where they already had an architectural style of their own. Moreover, they hardly wished to please the infidels. In fact, Muslim conquerors tore down Christian buildings and built their own places of worship on the sites as the Euro-Christians had done with the people they conquered. The Temple of the Rock in Jerusalem is an example. Also when the Turks conquered Christian Constantinople, they began to erect mosques in many places. One example was the cathedral of St. Sofia which was converted to a mosque after the Christian statuary was replaced with Islamic writing on the walls.

In the 18th and 19th centuries, as Euroman's interest in religion declined, new styles of architecture arose. These were secular buildings developed for use in Euro-commerce. Religion was out and business was in. And as the commercial interests of Euroman spread, so did his new building style. First came storehouses, then factories, utilitarian buildings of no esthetic interest. These were followed by office buildings and hotels. The artistic components in these buildings were minimal since their function was to make money, not to foster religious belief or fulfill esthetic needs. Only the international hotels used some degree of decoration since they had to attract high-paying customers. However, the decorations usually were superficial. In a few instances, the Euro-colonials tried to mimic local architectural styles. The British erected pseudo-Islamic-style public buildings, stations, and hotels in India and Malaysia.

In any event, as the great new commercial cities of the world grew, construction was almost exclusively in this Euro-commercial style, it being taken over by the newly independent nations as colonialism waned. In the end, great commercial centers like Bangkok, Tokyo, and Bombay were built almost exclusively in the new Euro-style, with small islands of traditional style buildings, usually religious, tucked away here and there. In places where there had not been any significant traditional buildings, and frequently not even any cities before the Euro-era began, as in the cities of sub-Saharan Africa, all buildings were of Euro-commercial style. In North and South America and the oceanic islands, including Australia, native construction styles were totally replaced as were almost all other aspects of culture. And in the end the great cities of the world became essentially the same clusters of tall, rectangular office buildings, apartments, hotels, factories, and storehouses surrounded by residential areas which ranged from posh suburbs to shanty-towns.

Although *sapiens* was primarily a seeing animal, he did have a fairly good sense of hearing. Thus, from earliest times he tried to make pleasing sounds, both directly with his voice and indirectly with instruments. And when Euroman encountered the various peoples of the world, he was exposed to their singing, playing, and dancing rhythms.

The music and dancing of tribals provided some diversion for Euroman after his hard day of keeping them in line. In fact, one of the few tribal activities that continued up to Takeover was the native dance spectacle. Popular documentary films usually included happy, dancing natives. However, these dances had changed a great deal from those of the days of native independence. The North American Indians, for instance, developed a pan-Indian style in the reservation period, designed to attract Indians from other tribes, as well as Euros. The old traditional styles like the Sundance and Ghost Dance by then had been abandoned or modified beyond recognition. These and other Indian styles which the Euros considered too vulgar or dangerous (frenzied dances for freedom) had been strongly discouraged or prohibited. The main pow-wow style consisted of a stomping shuffle to the accompaniment of rattles and drums, with the dancers garbed in modified Plains Indian garb.

The great civilizations had their own complex musical styles which continued despite Euro-influence, even though weakened.

Euroman never did get too concerned about the music of peoples he dominated unless it was combined with dances or rituals which threatened his dominance or seemed immoral to him. He, of course, never found war dances interesting since more often than not the war the natives were building themselves up to was against him. The Ghost Dance of the North American Indians particularly bothered him since in the final version it was claimed that Euroman would be eliminated. One of the last great performances resulted in the massacre at the Battle of Wounded Knee, followed by a prohibition of the ritual. Euroman's special double-speak usually referred to a massacre of Indians as a "battle." Once the natives were fully pacified, war dances could be performed again, and Euroman would even attend as a spectator.

Euroman had his own musical styles, the most prestigious being based on harmony and tone, usually involving large combinations of wind, stringed, and percussion instruments. In the early days of Euro-influence such orchestras were largely devoted to the production of religious scores. But through the 400 plus years of Euro-dominance, that music became secularized, just as did architecture, family life, medicine. science. and other aspects of Euro-life. By the end of the 20th century this style of music, known as classical, was played in symphonic form. The number of musicians in the orchestras ranged from dozens to hundreds. A variant of classical music was opera wherein singers acted out simple stories to the accompaniment of complex music.

Classical music was taken to cities everywhere that Euroman went and became established alongside the traditional art styles just as impressionistic works had been. It was taught in schools of higher learning in the non-Euro-world, and many members of the middle and upper classes, the educated, became accomplished in the new style. Quite a few became orchestra conductors or exceptional players..

In the meantime, the masters of the traditional styles in the Arabic world and South and East Asia kept traditional styles alive, if not flourishing.

Euroman also produced popular music, much less complicated pieces that were usually sung. As in opera, a main theme was romantic love, the singer either extolling the qualities of his/her beloved or bemoaning his/her loss. There were a variety of styles of popular music though the main one at the end of the

20th century was called rock and roll. It had the same general themes as other kinds of pop music, but was practically always sung in a very emotional or physical style by male singers, accompanying themselves on a guitar. Rock music was exported widely to the young colored people of the world. It went practically everywhere the ubiquitous blue jean went. The young of the colored peoples' world got their guitars and worked out their combos, first playing the ditties they had learned from Euro-players and later making up some of their own. The primary Euro-culture that produced such music was the United States.

Rock music was closely identified with the young and became a source of conflict between generations. This conflict was intensified by the fact that the musical style came at a time when the Euro-family had gone quite individualistic. The older Europeans felt more comfortable with earlier forms of music. In the world of the colored man the split in generations was more often between native traditional and rock.

One other significant form of music evolved in the Euroworld, jazz. This was a style which was highly rhythmic, improvisational, and played by smaller groups, both with and without singing. It was originally a music of American blacks. During their slave period the blacks were deprived of almost all their African heritage, and seemed to have tried to compensate most with music and the religion of the Euro-masters, Christianity. In any event, they became quite well known for their religious musical style which while continuing after the end of slavery in its religious aspect, also evolved into a secular style which came to be known as jazz. Religious themes were dropped and secular, popular themes such as romantic love were incorporated. Jazz spread to American whites first, then to other Euro-countries, and finally to cities of the colored man all over the world.

Euroman was instrumental in the dissemination of a number of other cultural traits which he had either invented or taken from the cultures of some colored men. Procedures for counting were all Euro-derived. Although all peoples had some method for calculating, the decimal system of Euroman was probably the most common, believed to have been derived from using the ten fingers. However, there also were some systems based on twenty, perhaps derived from a combination of fingers and toes. The English method of counting in twelves seems to have been quite unusual and created a conflict of numeral notation in the Euro-world. There was no doubt, however, that the decimal system used by other Euros was the best. For some very simple cultures there were systems of counting several only, then figuring the rest as a bunch. The Euro-method of using place numerals and the zero for nothing or multiples of ten (10, 100, 1,000) was called Arabic, though it is believed to have come from ancient India. In any event, Euroman disseminated this numerical system to the rest of the world so that by Takeover the one written symbol that could be read all over the world was the number.

Much the same happened with the calendar. Most people used suntime (360 plus days), moontime (30 days) and daytime (24 hours), all natural units derived from the motion of the heavenly bodies. However, the other units, hours, minutes, and seconds, were arbitrary and became universal only because the clock, a Euro-device, was so calibrated. So the world's travel, communication systems, and means of setting appointments was Euro-derived.

Maps, the charts that *sapiens* used to locate himself for travel were vital during the era of Euro-exploration. And as Euroman continued to travel and discover new places, he continually updated his maps. Moreover, since the center of his universe was in the northern hemisphere, all maps were oriented with north at the top. Also, topography was delineated with a series of concentric circles, the east-west orientation beginning in Europe (Longitude). So when the colored peoples got to travelling around, they used Euroman's maps. China ceased being the "middle kingdom" and the Himalayas became only the highest mountain range in the world, not the abode of the gods. All the mini-geographies of the tribals were absorbed by the new immensity of the world. What Zuni who had knowledge of the rest of the world could still honestly believe that 1,000-foot Thunder Mountain was a central, sacred place or what Apache could claim that the 8,000-foot Chiricahuas was the abode of the gods.

One might think that some customs would remain untouched in the days of Euroman, such as personal names. But for better or worse, not even that happened. One change that took place all over was the use of names for personal identification. In other cultures the name more often served as a means of establishing power or relationship to the supernatural. A person was vulnerable or powerful through his/her name. Thus, in many places either the given name was not freely divulged or one had a public name and a private one. Babies frequently were not named for several days or weeks until it seemed clear they would survive and were beyond the evil glances of envious visitors.

In some cases, clan membership was what counted, not the given name. One was a Lee or a Wong or a MacNeish or a MacArthur. Or one was simply known by one's occupation, an Archer or a Wheelwright or a Potter, as in England during the pre-Euro era. The given name was inconsequential. Later on some people would simply add son to the occupational or other family name to come up with Masterson or Sanderson or Ericson.

When modern states evolved and it was decreed imperative to keep track of citizens for purposes of taxation, conscription, licensing or other official reasons, such sloppy naming procedures would hardly do. One had to have a last name, a given name, and in many places a middle name. Of course, later on, as computer numbering became ever more significant, numbers began to replace names. Well before Takeover, though one still had to sign one's check, it would rarely be cashed unless a driver's license number was added. In the United States the social security number also was frequently used for this purpose.

A way that Euro-names became widely spread was through conversion to Christianity. Whenever missionaries lined up some natives to sprinkle water on them in the ritual of Baptism, a Christian first name was also given to the ex-pagan. So we got millions of John Rampaths and Christopher Wongs, Catherine Redwings, and Helen Cleareyes.

Another way was to simply give a new member of a society a name in the dominant language. This was particularly true of the United States during immigration from Europe. The immigration officials frequently found the names of European immigrants difficult to pronounce so they simply shortened them or gave them completely new names which they then put on the documents. Thus Przewalski became Walsh or Jones and Franciscus Shoboly Nagy became Frank Nagy.

The names of African slaves were changed totally. They had come from many tribes of the west coast of Africa with their own names in their own languages. To keep track of the new workers the slave owners simply gave them Euro-names, both first and last.

When the black-consciousness movement gained momentum in the 20th century, some blacks rejected their Euro-names as baggage from their slave past. One group, the Black Muslims, who claimed that their true cultural roots were from Islam, simply called themselves "X." Thus, they became Mary X or Peter X. Other blacks took names from whatever African language they decided was most appropriate. They became Jane Bambuti or Peter Chimboya.

One of the last indications that another people had lived in an area was the survival of the place names. Thus, apart from the written histories, the only indication that Indians lived in many parts of the United States was from the names they left behind: Chicago, Milwaukee, Alabama, Natchez, Malibu. On the other hand, there was New Amsterdam which was changed to New York. Others were Virginia, Williamsburg, and my all time favorite, the name of my home town, Indianapolis, and the state, Indiana. To my knowledge no descendants of Indians were living there. Indiana was one of the many places east of the Mississippi that had solved its Indian problem early. Evidently though, according to the names, there must have been quite a few around at one time. Mexico, of course, has many places with Aztec names as Peru has Inca names, counterbalanced by Spanish.

In all cultures *sapiens* had specific ways of greeting others. At the very least one needed to indicate that one's intentions were friendly, if they were. If they were not, one did not bother with greetings or other niceties; one just attacked. Naturally, these greetings varied from place to place. Far Easterners bowed, which also indicated deference. The deeper the bow, the deeper the homage being paid to a person of greater prestige. During the days of Oriental royalty, even greater obeisance was shown; the lower ranking person crawled on his knees, sometimes touching the foot of the higher ranked. The Hindu Buddhist world developed a kind of prayer-greeting which consisted of folding the hands and gesturing toward the greetee. The Arabs presented the right hand with the palm up along with an appropriate phrase.

When they were spreading throughout the-world, the Euros were coming out of a time called the Medieval period. This was a time "when knighthood was in flower," as they used to say. The enforcers were armed horsemen called knights and the "enforcees" were called peasants. The job of the knights was to keep the peasants in line, as well as protect their masters' territory. They were evidently "fast on the draw" as was said later of the gunfighter in the Old West. If the knights were displeased, out came the trusty blade, the sword. If they were pleased, they presented their paw, their sword hand, minus the sword. It is said that the Euro-greeting, the handshake, came from this knightly gesture of peace.

So the Euros took this gesture wherever they went and the locals came to expect it of them. It evolved into the international gesture of friendly intentions, although it could be less than sincere in many instances. The symbolic nature of clasped hands was so deeply entrenched in the United States, it was stencilled on the crates sent to colored men's lands whether they contained arms for making war or economic assistance.

Then there were drugs. *Sapiens* learned that some substances, especially from plants that he took by mouth, and later in other ways, could alter his perception of the world. These drugs would do at least three things to his body and mind that he found interesting: get it excited (uppers), calm it down (downers) or give him visions (hallucinogens).

The Euros were not much in inventing their own, but they were certainly good at distributing drugs they learned about from others. The only real mind-altering drug they had when their age of expansion began, was alcohol, in wine, an ancient fermented drink, and liquor, distilled fermented juice invented by the Arabs. Both were carried everywhere by the Euros and almost the entire world imbibed. Fermented drinks were already known in most parts of Asia and Africa, but many tribals in the Americas and Australia did not have them. Distilled alcohol was an innovation almost everywhere. In any event, alcoholic consumption was spread all over, and became one of the chief consolations of tribals when their culture was being destroyed and their land expropriated by Euros.

The Euros adopted coffee from East Africans, tea from the Chinese, and tobacco from the American Indians. And just as with alcohol, they spread these substances everywhere they went. Coffee and tea were both uppers and became morning drinks of the world as well as on social occasions. Presumably people conversed better with one of these drinks or alcohol in hand.

Tobacco was used to calm one down or probably just as often to dispel boredom. And although the Indians, from whom Euroman took over the weed, used it in several ways, the way he took it turned out to be damaging to the health. Tobacco smoking was believed by most to be the primary cause of lung cancer in the late 20th century, truly "the red man's revenge."

A mild downer, marijuana, was also spread widely by Euroman. When he conquered the Andean Indians they were using an upper from the leaf of the coca plant. Euroman experimented with the substance and decided to make it stronger by concentrating it. The new drug, cocaine, became one of the mainstays of the international drug market. Another was from South Asia, opium. The drug from this plant was a super downer, deliberately spread by colonial Euroman at first and later distributed by underworld types. So among other roles, Euroman became the dream merchant without parallel.

14

The End of an Era

As I stood before the board, chalk in hand, I realized how repetitious I was getting. But I had an explanation. I had not been a *sapiens* for well nigh 100 years without learning the noble art of explanation. This was the third time I had turned to the board, though perhaps the last, I thought.

Actually I had two explanations. First was that I had been deprived of the use of this kind of exposition for about 20 years since I had retired. How I had managed so long without teaching and without putting my thoughts on a blackboard I had difficulty imagining. And then by a strange quirk of fate I had been brought back into lecturing through the Atierran draft! I was unexpectedly teaching the last class of *sapiens;* I was doing the last ethnography.

Also it was comfortable to note that this was the third time I was returning to the board. I was using the primary sacred number of Euro-culture. The third time would be the charm, I mused.

Anyway, there I stood, my ideas about the last lecture welling up in my mind. I felt a sense of accomplishment, as well as one of sadness. Would this be all? Were there to be no more words about the culture of Euroman? Would the history of *sapiens* end after this? In any event, it was consoling that it would have been put down, and anthropology, as I understood it, would have been served.

I had been through enough sessions by this time to suspect that Mary listened in and watched what I was doing even before she turned on the lights. I had noticed a very small light in the lower left corner of her screen when the main lights were on and she was talking, although I had never seen it come on. I suspected it indicated that the audio-video circuits were functioning, whether the screen was lit or not, and that it indicated eavesdropping.

Interesting, I thought, the idea of gathering information secretly. There had been much rhetoric criticizing the practice in my profession even though some occupations, particularly in government, depended almost completely on gathering information secretly. The entire profession of espionage depended on secret information. But there were other professions, for instance, medical nursing. In it there were no elaborate techniques of surveillance as in government espionage; it was nothing more complex than watching surreptitiously while seeming to be doing something else, say making the bed or dressing a wound. The nurse would be checking to see if the patient had complied with instructions, or what they called otherwise, "doctors orders." Patients frequently did not comply because they either didn't understand why they were supposed to do something or they disagreed with the medical authorities.

I had become familiar with the nurse's penchant for checking from the actions of my wife, a registered nurse. Jeanne had been a dedicated checker since I first knew her. I had become thoroughly accustomed to her flicking eyeballs when she stole a sideways glance at my doings. Sometimes she wouldn't get her eyeballs back to her magazine or the TV set quickly enough. I knew it was a general practice of nurses checking patients. It was perhaps inevitable that she would do this with her spouse since what he did was more significant to her. A thought crossed my mind: Was Mary an ex-nurse?"

I immediately felt very foolish. Was I flipping my lid? Had I been in this room, spilling my guts, too long? Mary was a computer. How could she have been a nurse? I tried to straighten out my thoughts. I was honestly perturbed. Admittedly the whole affair had been bizarre, but somehow I had imagined myself as a master of adjustment. If the pre-conquest Indians had survived the rigors of total defeat and almost complete cultural transformation, and could still relate the old way of life to anthropological investigators, as a trained social scientist, I should be able to do the same to a friendly computer. Anyway, I soon settled down, ready to deal with electronic Mary again.

When she still failed to come on, I turned my attention back to the board. First, I tried a stick figure, and then put a variety of hats on his head. I finally ended up with two which were better than the previous ones.

After I had studied the two figures a while, the lights came on. Then, "Hi, Pete, still busy?"

"Hi Mary. Yup. Trying to make images of my thoughts."

"I see. This time it's stick figures wearing headdresses? Must be an explanation for that, knowing how you operate."

"There is, of course, even if the artistic level is not so great. But I never was very good at putting my ideas into images. As I told you before, I was always a word man."

"Well anyway, they are figures that look like some sort of humanoid. Sort of primitive art?"

"Yes, quite a few simple societies developed stick figures to portray their ideas. Perhaps one of the oldest and most widespread art styles was of stick figures."

"But yours have headdresses?"

"Yes. The figures were simple enough to do but the hats, oops headdresses, were difficult, both to decide on and to draw. If I borrowed the idea of stick figures from primitive art, it is mainly because I wasn't much good at making other kinds of figures, and they served well enough. I was mainly interested in portraying the headdresses. This idea struck me as I was musing on my role in telling this history which I think of as twofold."

"You mean there is something about you and two hats?"

"That's it, Mary. During most of my life I wore two hats, as the saying goes, and certainly for this account."

"You'll have to be more explicit, Pete. What does wearing two hats have to do with the history you have been relating?"

"Well, you may know that we *sapiens* as individuals were not always consistent. I suspect other animals were not either, but here we will stick to people. So we worked out a sort of solution by splitting ourselves in two or more parts or roles. I don't know if there is a name for this, but there ought to be. We did have a name for a psychosis of this sort. It was called schizophrenia. As was typical of medicine, we had to invent the longest word possible to describe the condition. Anyway, I split up in my very early professional days, and I guess I've been split ever since. In other words, I wore two hats, which means really that I had two ways of looking at the same occurrences." Then I quickly added, "I suspect many anthropologists did. The likelihood was built into the profession."

"Okay, go on Pete. Sounds like this might have some relevance for analyzing your document."

"I think it does. Anyway, here goes. You see, sometime after I became maladjusted in my own culture and got involved in anthropological studies, I looked to other cultures as possibly being better. And whether they were or not, I wanted them to be. That's why I was usually rooting for the tribal or the peasant. I really wanted them to have a fair shake. I was then wearing the hat/headdress of a caring, cross-culturalist. What does such a hat look like? I'm not sure, Mar, but perhaps it was something like a tribal would wear, say a feather headdress. Needless to say, Mar, wanting something to be the case was hardly a scientific attitude. Many of the wrong turns of science occurred because people justified theories they wanted to be true. But then I continued my studies and learned that such ideas were only wishful thinking. There was no way the American Indian or any other tribal, or any peasant villager for that matter, was going to get a fair deal, much less have his old way of life restored. We already had a theory to explain why that wouldn't happen. It was called the Theory of Evolution, or more specifically, Natural Selection, and though it had originally been worked out to explain how biological species replaced each other, it was just as appropriate to explain cultural competition and the replacement of one culture by another. The reality was that the weaker were replaced by the stronger. And by stronger I do not mean fang and claw, I mean any characteristics which gave a species or culture an advantage in competition. Most often with *sapiens* this was some tool, say a spear or firearm, but as you will see shortly, it could also be a social or religious force. And to make a long story short, any species or culture which could overrun or remold another was a winner in the evolutionary sweepstakes. Thus 'takeovers' were natural in the scheme of things."

"Wow, Pete. You are giving me your theory of survival. That's what you call your other hat, I take it."

"Yeah, 'fraid so, Mary. This hat is the scientist's, the professional trained to look at events without emotion, irrespective of their rightness or wrongness, their morality or immorality. One was supposed to look at the world as it was, not as one wished it to be. This was hard, but I think I learned how to do it. Then I could see that 'takeovers' were as inevitable as breathing."

"And what was the hat, Pete? Just out of curiosity." I see your drawing but I must confess it is not clear."

I looked at my second stick figure on the chalkboard and it still bothered me. It certainly wasn't very good. I sighed. "It is a mortarboard, the headcovering people wore when they graduated from college and when they were brought back to the graduation ceremonies later in life to receive honorary degrees. It was a custom derived from our Medieval period."

"Hmmm. I think I do remember seeing that headcovering in some photos, probably of a graduation ceremony."

"That's it, Mar. The Euros tried to make a big deal out of completing university studies even though our students were not nearly as accomplished as the

Orientals. The Euros called it "commencement," implying I suppose that the important events of the person's life were just beginning. But at a standard graduation there also were a few people who were judged to have accomplished something important on their own. They were given honorary degrees."

"So Pete, you were thinking you were wearing the hat of a recipient of an honorary degree? That you had accomplished something important?"

I laughed. "No Mar, not me. I was never likely to be rewarded as a great achiever, both because of my profession, which was offbeat, and because I was a heretic in my own field. The people who received honorary degrees were those who achieved in highly respected professions and also in directions which were approved by people in their field and by other social leaders. Thus it was normal for an innovative surgeon or a person who worked for peace to receive this recognition. You see Euro-cultural leaders were much in favor of curing illness and making peace even though the cultures created a lot of illness (mental and diet related problems particularly) and developed the most bloody wars of history. Anyway we gave degrees to people who did something in those fields, normally without doing anything to change the cultural patterns which were causing them. I never heard of an anthropologist who was even considered. In general the social sciences were not very prestigious in the Euro-world."

"So why do you put this mortarboard on your other self, Pete?"

"I'm simply using it as a symbol of the scientific method, even if we anthropologists were not considered to be as important as the hard scientists, the physicists and chemists. But even a social scientist would be more likely to have a naturalistic outlook than someone with another kind of headpiece, say a baseball cap or a felt hat. It's not a perfect symbol, I admit, but I don't know what other kind of headpiece to put on a scientist. The "mad" scientist was typically thought of as a wild-eyed zealot whose hair went in all directions, like Albert Einstein. But I am only trying to indicate that I harbored this kind of view up to my middle years. To me it explained much of what was going on in the world even if it sometimes offended the person I was with who was wearing the other hat, the one of the caring, cross-culturalist."

"It sounds like a heavy burden, Pete, having to keep two contradictory viewpoints. Don't you agree?"

"Well, it was one of the most difficult things I had to resolve in life. But I managed as did most of my colleagues, many of whom I'm sure had the same problem; although a few took drastic measures like turning to alcohol or committing suicide. Sometimes, I thought we *sapiens* had a built-in capacity for living with contradictions, sort of like for learning language.

Mary's screen went dimmer, then came on with "Transition."

I figured she was getting tired of my two-hat analogy and wanted to get on with it. I sort of felt I'd said enough about that, so I put the chalk on the scooped-out place below the blackboard and sat down.

She started in again. "Okay Pete, I got it, I think. And I take it you plan to be your mortarboard self in this section, that is, your scientific self?"

"I think I will, even though it is still a little hard for me to abandon my feather headdress. Believe it or not, I really cared for those tribals and peasants."

"I had noticed your bias in favor of them in the previous accounts, and even of the old civilizations when they had to deal with the Euro-onslaught. But now you will look at the same events dispassionately, eh?"

A troubling idea came to me and I decided to get it on record. "First, and though I hate to add another caveat, a legalese term that I practically never have a chance to use, Mar, but there is one. The hat situation is even more complicated than I indicated. Instead of wearing two hats, I probably wear three. The third is that of the standard culture carrier, a person who had been properly brainwashed. This results in a total commitment to the culture. The majority of mankind undoubtedly have been this way, making up the backbone of any well-functioning culture. Their attitude normally would be 'my culture, right or wrong.' And when there were inevitable conflicts about the rightness of certain actions, there were always cultural rationalizations to justify the actions taken. Thus, most Americans believed it was the natural destiny for Euroman to displace the Indians, and most Germans in the Nazi era thought it was proper for their kind to get rid of Jews.

"Anyway, I see it as the billed cap, what the all-American working male or sports enthusiast would wear. And needless to say, anthropologists as a group did not often wear that hat. If anything, they wore the feather headdress of 'their people' more often."

"But I'll try to be dispassionate, Mar. It is somewhat easier now to wear the different hats since even before Takeover, the Euro-onslaught was over and a shift in dominance was well on its way. We were into a transition when you Atierrans came. The old Euro-era was over and new powers were rapidly gaining momentum."

"So that's it, Pete? We are left to discuss the transition or the change in dominance?"

"That's what we'll do now. And then probably I should stop because that would be the end of history as we knew it. " I added however, "I have thought of taking one more day to talk about the odds and ends of our ethnography which are still not resolved, what would be considered the wrap-up. I particularly want to discuss what history is all about. You see, in my lifetime I changed my mind about it almost totally. When I was young and learned my history in school, I accepted the idea that it was the study of what really happened in the past. As I grew older and learned more about the past, I gradually came to the opinion that history was a study of modified accounts and myths designed to support the existing culture. It only partly reflected what really had gone on."

"That does sound interesting, though I'm not sure I can get another day for us. I know Central already has another program booked." Pause. "But maybe I can have it postponed. I'll contact Central and look into it."

"You could also ask me some questions about matters you feel need clarification or amplification. The old ethnographers who studied the tribal cultures used to do that after they looked over their notes."

"Doesn't sound bad. But for now shall we go on with our discussion of the transition? And you might do your usual, explain what occurred before the Euro-onslaught."

"Well, we know this mostly from the early history of civilizations that had writing, which was backed up by some little information from archeology. All anthropologists agreed that man began as a nomadic hunter and gatherer, a cultural legacy of his primate ancestry. Change was constant but slow for the next three million years. Then in parts of the world man learned how to domesticate some of the animals and plants. This new food resource gave the farming peoples many advantages over the nomadic tribes, including a more reliable food source, larger populations, and more stability, and with an increased hunger for land. Evolutionary forces went into effect as man the cultivator spread into the territory of the gatherer. He then evolved into man the city-builder, who was even more land hungry. And by this time he had learned how to organize the peasants into armies. This occurred simultaneously in many places. Urban civilizations began displacing or absorbing simple cultivators as well as nomadic gatherers. The simple cultures that survived did so primarily in marginal places urban man did not then want or had not yet reached. Thus, the major regions left to the hunter-gatherers by 1492 were Australia, some pockets in Africa and Asia, and the northern and southern parts of the Americas. The simple cultivators retained somewhat more territory. Their main concentrations were the islands of the Pacific, sub-Saharan Africa, the temperate zone of North America, and the lowland tropics of South America. Elsewhere, nation states reigned. These were also referred to as civilizations which not only kept moving into the territory of the hunter-gatherers and simple cultivators, but began competing with and replacing one another. Some of these were Imperial China, India, Egypt, Assyria, Persia, Greece, Rome, and expansionistic cultures in Mexico and Peru. Thus, through automatic evolutionary processes, the domain of the hunter-gatherer and simple cultivator had shrunk a great deal by 1492. In the next 500 years it was to shrink to almost nothing, in no small part due to the spread of Euroman."

We have already gone over the history of Euroman in considerable detail. He conquered everywhere he went, subduing and controlling great civilizations as well as gobbling up the remaining gatherers and simple cultivators even faster than earlier urban cultures had done. In the late 1930s Euro-industrial and military power were at their peak, and the Euros were on the brink of a second major war. Only one of the combatants in the coming conflict, Germany, had a Euro-culture. The other, Japan, in its war-making mode, was a direct consequence of Euro-influence. A hundred years before, the Japanese had tried to seal their borders from Euro-intrusion, but were forced by American military might to "open up." From then on they industrialized and militarized, using Euro-models and techniques. They had fought before, but only among themselves, with close neighbors and according to their own social principles. Their military ventures in the 19th and 20th centuries were strictly Euro-style. Even their "East Asian Co-prosperity Sphere" of W.W. II was a latter-day version of Euro-colonialism. In a sense modern Japan was Euro's Frankenstein, a monster of power which got out of control.

Other cracks were appearing in the Euro-structure, though not yet obviously. The colonial world was becoming restive. Many colored men had gone to the universities of their colonial masters and had learned to their surprise that the Euros proclaimed self-determination and freedom, even if under each's respective ideological umbrella. Some leaders of colored men's countries were not even sure they wanted to fight in the new war. In India, for instance, the second major independence movement evolved. But the native leaders were willing to send their peasant soldiers to kill and be killed only if India also received its independence as part of the package. Mahatma Gandhi became a big name in the Euro-world in this regard.

Then the war broke out in earnest and the Euros organized probably their last major effort to maintain dominance in the Pacific area. In Europe the Euros were fighting each other, along with the peasant soldiers of their colonies. The Anglo-Americans and their allies won the war. It seemed that they, and particularly the prime victors, the Americans, were as powerful as ever.

After the war, however, changes took place rapidly, most of which weakened the Euros. The colonial world rapidly broke up, partly because the colored men of the world had seen clearly that the Euros were not invincible even if they had won the war. Also the colonies had become increasingly expensive to maintain. No longer was there much worth looting, or good dumping grounds for manufactured goods. Further, unexpected competition emerged. After W.W. II the nations of the colored men got started in industry. They had learned an even more important lesson than the need for independence, the need to take over Euro-technology.

Anthropology had long taught that the basic process of change at all cultural levels was the diffusion of innovations. Although it was comforting for nationalists to emphasize inventions as the bases for their progress, and though such did have to occur, once they did, a new way of accomplishing a goal was easy to borrow. It made no sense to reinvent the wheel. Thus the great cultures of the world were those which borrowed heaviest. The Aztecs and Inca had borrowed wholesale from the previous civilizations as had the Romans and Greeks. Both England and America practiced such in their days of glory. They had borrowed from everywhere. And Japan, despite its insularity and efforts to maintain its own integrity by turning inward, had already gone through extensive borrowing. It had taken over many customs from the Chinese, including the writing system, Buddhism, some Confucianism, porcelain, the tea ceremony, decorative carp (koi), dwarfed trees (bonsai), the bow as a greeting, artistic and architectural styles, and much else.

The initial burst of industrialization after the Meiji Reformation in the 19th century stemmed from a deliberate decision to borrow Euro-technology. Its very success had sponsored the building of a Euro-style war machine with which Euro-style military expansion was possible. Except for some piracy, for the first time in their history the Japanese left their home islands. Up to the time of industrialization they had been content to expand in Japan, using their energy to displace the Ainu, their tribals.

But their industrialization became more and more efficient until a war machine was built which could challenge the great Euro-powers, first the Russians then the United States, as well as the other colonial powers.

In retrospect, and despite the suffering they went through, probably the best thing that could have happened to the Japanese, did. They lost the last war. The new age would be one of pure mercantilism. It would no longer be necessary to grab other peoples' land or subjugate them in order to dominate them. One would be able to buy and sell them into submission.

The victors of W.W.II. did not recognize the inner strength and resilience of Japanese culture. They established a Pax Americana to prevent further military opposition. The front man and major architect of the peace conditions was General Douglas MacArthur who after having the top military leaders hanged (a normal procedure by victors), took away some symbolic power from the emperor and forbade any further military buildup. Further, he set up conditions for a capitalist democracy with unadulterated ethnocentrism. But since the Americans were as much in favor of industrialization as of apple pie and hamburgers, they left that sector of Japanese culture untouched. So the Japanese trundled out their wheelbarrows, picked up the rubble, and converted their wartime factories to make cloth, motorcycles, motor vehicles, cameras, electronic equipment, etc.

Further, they had learned a valuable lesson from their previous efforts to make goods for export. In general the Japanese were good learners. Before W.W. II the goods they had made emphasized cheapness. When I was a boy in the '30s we always spoke snidely of "Japanese junk," cheap goods available at the "five and dime stores." Ironically, the primary symbol of quality then was the label "Made in America."

When the Japanese tooled up their factories after W.W.II, they included quality control. They learned much of this from an American who had not been heeded in the U.S. By then the Americans were moving into a kind of production which had "built-in obsolescence." But it really doesn't matter who the Japanese learned it from. They got the idea and it helped to create the "Japanese miracle." It was apparent by the end of the 20th century that the super-producer of the world had become Japan. They also became the international financiers and the primary investors.

Though leaders of dominant cultures had never been very good at predicting the future of societies, military men were probably less well equipped than civilians. They had been trained to organize military machines to defeat other such machines. Other aspects of culture were not generally part of that training. But because they did move into positions of power from having defeated others or simply intimidating them, militarists were often given assignments far outside their expertise.

I have sometimes fantasized a conversation between the spirits of Commodore Matthew Perry and General Douglas MacArthur, the two military leaders most responsible historically for creating the Japanese miracle.

MacArthur, fingering his decorated, billed officer's cap: "I just don't understand how they managed to challenge us so quickly when they had been beaten to their knees and had to start from scratch."

Perry: "Well now, not completely from scratch. There was a lot of knowledge around from their previous effort to modernize. And if you think the industrial achievement after W.W.II was a miracle, think of what happened after the "opening." When we steamed into Yokohama Harbor with our letter of demand that they open their ports to foreign trade, they were living in the Middle Ages. They called their lords shoguns and each one had established a little dynasty through the services of their soldier caste, the samurai. They didn't have a single warship and still depended primarily on swords in land warfare. And true industry just didn't exist. Imagine then how we navy people felt when after the turnaround we saw the Japanese invade Korea, then Russia, then China, and finally the United States and all other countries that resisted their expansion in Asia. Further, they were using weaponry by that time that any Western army or navy would have been proud of. It's an outcome which is dreary to contemplate."

MacArthur, silent for a moment, then with a sigh: "We saw to it that they couldn't rebuild their war machine after W.W.II. But even that seemed to work against us. By the end of the 20th century the Japanese were literally flooding us and the rest of the world with high-quality manufactured goods. And because of the treaty we'd imposed on them, they had no major military establishment to support. The soldiers we sent to defend Japan could barely afford to pay for their goods except on the military bases." Pause, then fumbling in his pockets he produces a corncob pipe: "But then what else could we have done? We certainly couldn't have taken a chance on them building up their war machine again." Starting to fill the pipe, "You know, Commodore, I can tell you because you know military matters, but there are few civilians, alive or in the spirit world, that I would say this to. The real fact of the matter was that we barely defeated them. I think man for man their soldiers were better than ours. And so far as military tactics were concerned, no one could fault the Nipponese. As I see it, the main reason we defeated them was our industrial might. We simply built our war machines faster than they could destroy them." Hunched down, still cradling his corncob. "It is depressing, Commodore."

Perry: "And ironically they are now the most highly industrialized nation on earth." He sipped from a wine glass of pale liquid, soft focussed in spirit light. "But to go back to the early days, who could have guessed when we demanded that they open their ports that they would become such formidable industrialists themselves? After all, we only wanted a free market because our factories in New England were expanding so rapidly." He paused, looking thoughtful. "And we certainly couldn't have forbidden them to set up factories themselves. But so fast!" He shook his head.

MacArthur, striking a long kitchen match and applying it to his corncob, the flame flashing in and out of the bowl several times before he blows out a little cloud of smoke: "Okay Matthew, let's admit it, you and I and the government wiseacres who developed the policies certainly made a couple of real boo-

boos. In my day they would have called them lulus. But what has bothered me even more is trying to figure out the characteristics which made them so formidable. I've thought about this for years. After all, here in the spirit world, time is what we have in abundance and now I seem to be getting a clue. When we went to school right on through college we were taught that our kind, white men, were the greatest achievers worldwide, that we were the most moral people and that we had attained greater individual freedom than anyone else before us. And though I don't want to overemphasize race as an explanation, we were taught that the white man was probably more fit." Gazing at the slowly eddying smoke from his corncob, he goes on, "We were quite big men physically, you know, especially compared to the Japanese. Even the Chinese called them bandy legged dwarfs in the early days. And I remember that though one didn't see many Japanese prisoners in W.W. II (against the code of bushido), when one did, they looked positively puny standing next to their American marine guards." He paused again, stared into space, puffed slowly on his pipe, then, "And you know, Matthew, we were also taught that our country was so rich in land and natural resources that we couldn't help being a rich nation. By comparison, the Japanese were among the poorest in the world in both land and natural resources. I believe their whole island chain was about as big as the state of California; and they had half as many people as the United States." MacArthur stopped then and frowned, seemingly at a loss for words from the depressing nature of his thoughts.

Patting MacArthur on the shoulder, the commodore replies: "I know, I know, Douglas. It boggles the mind. And whatever else one can deduce, one fact is clear, that our understanding of the Japanese as well as what makes a country rich certainly could have been better. Our analysts and policymakers were hardly brilliant." He sipped from his wine glass, then as an aside, "And another characteristic I've wondered about is their willingness to work so hard. I had always thought the Europeans, especially we Americans, were such hardworking people. Didn't you believe that, Douglas?"

"I sure did Matthew, in the early days. But in fact, if there was anything different about the Japanese that I could see, it was that they worked harder than we did. Why, after the dust settled from the Big War, those devils were out in no time, removing the rubble and rebuilding their cities. And before you knew it, their first primitive motorcycle was coming off the assembly line." Pause, eyes straight ahead, concentrating. "And we've both seen from here in Ghostland that they have hardly let up all these past decades. Only now do some seem to feel it's okay to take Saturday off."

I noticed then that Mary's light was blinking which surprised me. Before, she had always remained noncommunicative during my major delivery, simply leaving the letters REC on her screen. Anyway, I answered, thinking that she might be getting fed up with my hypothetical exchange. I said, "Hi Mar, something wrong?"

"No Pete. I've been listening with interest to the conversation between the commodore and the general. And the idea occurred to me that you were really wound up, that you had thought about the Japanese threat for a long time."

I felt relieved. "That's true. You know, like most Americans I suppose, I went through enormous changes in my opinion of the Japanese during my lifetime. When I first heard about them, I was a boy growing up in the midwest. They were vague Orientals who filled our low-priced stores with cheap junk. Then when I reached manhood they became our enemies by bombing our main naval base in the Pacific. Americans generally got very indignant about that, most not knowing or conveniently forgetting that we had taken it away from the native Hawaiians, along with a number of other pieces of real estate in the Pacific. The Japanese didn't get in on the island-grabbing binge in the 18th and l9th centuries, being too busy fending off the Euros and building their first international war machine."

"In a year or so after the Big War started, it became quite clear that the Japanese were not going to be a pushover. Then our industrial machine got cranked up and we started pushing them back. Then we devastated their cities with standard bombs and finally dropped our new weapon, the atomic bomb, on two of their cities. And we were the victors. It was always characteristic of *sapiens* to use weapons the other side did not have, no matter how destructive. Shortly before that we had defeated our errant brothers in Europe, the Germans. We Americans thought we were the greatest then and hardly considered others as industrial competitors."

"Then the colonial world shrank to nothingness, and while most of the European nations began to rebuild, the Americans indulged in a great orgy of consumerism. There were a few signs that change was in the wind, but Americans didn't really take it seriously through the sixties."

"One botheration was in the American auto industry. The designers in Detroit kept making bigger, more powerful, and more ornate machines and most American consumers bought them, as well as other luxury goods. They moved into the suburbs with their Detroit tanks and lived "the life of Riley.""

"Then the first industrial miracle occurred. Germany began to produce and export a tough little car designed in the Hitler Era, called the Volkswagen. Soon, these "beetles" were on all the roads, zipping in and out of the lines of Detroit behemoths. Both the German and Japanese made cars were gas sippers while the American cars were guzzlers. But at that time of affluence, only poor students and social oddballs cared about gas consumption. Besides, the Americans had the cheapest gas in the world because the government would not tax it heavily."

"At about this time the first Japanese motorcycles appeared on the American roads, soon to be followed by tough little cars similar to the 'beetle.' The Japanese 'miracle' was beginning."

"Americans slowly began to take the Japanese seriously. In the next thirty years, the industrial outpouring increased until by the 1990s the country was the super-giant in the world of industry and finance. They even took over much of the export industry of their fellow miracle workers, the Germans, coming to dominate the auto, camera, watch, and electronic industries among others. As one of the American admen would have said, 'They came a long way, baby."

"Yes, I was right. You certainly were caught up in these thoughts, Pete. And by the way, since you've thought about it so much, what do you think was the major difference between the Americans and Japanese as producers? Do you think General MacArthur's idea that they worked so hard explains it?"

"That was an obvious difference, Mar, but I do not find it the main reason. The Americans also used to have a reputation for working hard. Our frontier settlers certainly worked their butts off cutting down the forests, breaking the sod of the prairies, and otherwise ravaging the land; and the immigrants of the 19th and 20th centuries were no slouches in factories and small businesses."

"The Euros had a high-flung theory to explain it, the Protestant Ethic, developed by a sociologist, Herr Weber. He claimed that the Calvinists, a central sect of Protestants, believed that a person who succeeded in this life was the chosen of God. Success was the proof. So to put it bluntly, those people worked their asses off to achieve worldly success."

"And you have doubts about this theory?"

"Well now look, Mary. Apart from the Euros, the Japanese, as well as several other Orientals, were among the hardest working peoples the world had known, and believe you me there were mighty few Calvinists among the Japanese. That was obviously an ethnocentric explanation by a European, made when Europe was on top of the heap."

"Okay Pete. I can accept that, but knowing you, I'm sure you have a hypothesis of your own."

I grinned. Mary knew me. I went on. "In the first place, I don't doubt that the Japanese were hardworking, but I don't think that was the main reason for their achievement. The major difference, especially between the Americans and the Orientals generally, was in their respective emphasis on the individual versus the group. The 'Japanese way' was the group way. Americans even tried to inject some of 'their way' into American factories and businesses, without of course changing the value they placed on individualism. But the difference was too deep to be responsive to patchwork change.

"It went way back, Mar, I'm not sure how far on the Japanese side; but certainly back to the Medieval period, the Shogunate. Even then the Japanese had a tightly organized society in which each individual had a duty and a responsibility. For instance, a popular theme in their plays at the time, concerned masterless warriors (ronin) who wandered about in distress because they needed to serve a master in order to be content even if this required them to fight to the death. This same dedication to the master as the head of the group continued when the Japanese modernized. The soldier and sailor in the new Euro-style military machine served their officers with great loyalty as representatives of the supreme authority, the emperor. I suspect the new industrial workers were just as dedicated.

"Americans had a lot of trouble understanding this kind of dedication in W.W.II, particularly in respect to the willingness of the Japanese to commit suicide for the cause. Of course, the main reason this appeared bizarre to Americans was that they were generally ignorant of Japanese history. *Seppuku*, or what was more popularly known as hari-kiri, was an old tradition. Anyway, one aviation tactic which was followed in no other culture was *kamikaze*. A flyer had to dive bomb his plane with its explosives into an enemy warship. A type of minisubmarine was designed to perform the same function. Other cultures did sometimes promise great rewards to those killed in battle, but few asked their warriors to deliberately kill themselves in the process. To say the least, this was a difficult tactic to counteract, and it certainly took a lot of suppression of individuality on the part of the pilot. He was doing it for the group.

"Then too, the Japanese taught their soldiers the code of *bushido* which included the idea that no Japanese soldier could be taken prisoner. And, in general, they fought to the end. When the Japanese were losing the war, being driven out of one island after another, the American soldiers had to literally burn them out of caves with flame throwers.

"Then too, there were some spectacular mass suicides including women and children when it became obvious that the local troops were going to lose a battle.

"Another side effect of this code of commitment which Americans had difficulty in understanding was what the few soldiers who were captured did. As in any group of people, there were some who were out of step with their fellows. They surrendered rather than die. It was claimed by some that most of those who were taken were men who had been captured while bathing on the seashore and were without their weapons or clothes. In any event, there weren't many, but the ones that did surrender were not reluctant to report military information about their own units to their American captors. Anthropologists, who were working at the time to help the Allies understand Japanese war tactics, claimed that such behavior by a Japanese prisoner did not indicate he was a traitor in the American sense. Rather, since he was already dead according to the code of *bushido*, he logically had become one of the enemy, and any information he provided was being given to his own kind, the Americans.

"But perhaps the most powerful indication of commitment to the group came from the lone survivors on Pacific islands, some of whom continued for thirty or more years as members of the empire, long after it had been dismantled. Lone Japanese soldiers, cut off from their comrades in battle, hid out in jungle areas on Guam, the Philippines, and other Pacific islands. The battles had surged past them, and they found themselves alone in isolated jungle areas. And since they had never abandoned the code of *bushido*, and did not know the war had ended, they continued to act as soldiers. Their main job then was to survive, which they did, Robinson Crusoe-style, by building places to live and growing their own food. Some even contacted local natives.

"When the various island authorities learned of these jungle hideaways, they tried to let them know the war was over with leaflets, posters, and loudspeakers, and that the hideouts should come in to be repatriated to Japan. Quite a few refused to believe the messages were from their own people, the last giving himself up more than thirty years after the war ended."

"I agree, Pete. That is real commitment to the group. I take it this was not common in other cultures."

"I never heard of anything like it anywhere else. Generally, individuals who were cut off soon joined another social group, even when they thought it was dangerous. You may remember Ishi, Mar, the last wild Indian in North America. After the last of his own people were gone, he decided to become a Euro, no matter the danger."

"So how was all this commitment channelled after the war?"

"Well the emperor was declared to be nothing more than a normal mortal by the orders of General MacArthur. To the Americans one of the emperor's last great, and surprising acts of authority was to declare that the Japanese had surrendered and that all men should put down their arms. Because of the ferocity with which the Japanese soldier had fought during the war, the Americans expected massive resistance when they occupied the home islands, even though Japan had been thoroughly devastated by warplanes. But true to Japanese acceptance of authority, the entire population ceased fighting immediately and the occupation was totally peaceful."

"Wow, as you would say, Pete. That is impressive. And then as you mentioned before, the Japanese began the job of rebuilding?"

"Yes, they did, and under a freely elected government, also as decreed by MacArthur and company. Some claimed that the Japanese then viewed the General as the "white shogun." And with more or less the same dedication of the old days, they got busy in the factories and offices of the new industrial shogunate. To sum up, with all the dramatic changes in their history, there was this uninterrupted commitment to the group. And it moved mountains, so to speak.

"I can see that such dedication would be hard to compete with. It really did not depend on the existence of significant leaders. When one was competing, in war or industry, it was with the whole group. Even the emperor was no more than a symbol of the group."

"Yes, and it really went against the grain of the Americans, and of the other Euros to a lesser extent. You see, Americans valued the individual above all else. They claimed the government was 'for the people' not the opposite; though in actual fact they meant 'for the individual.' There were oddball types on occasion such as President Kennedy who gained a minor reputation for saying 'Ask not what the government can do for you, but what you can do for the government.' In general, however, most Americans thought that the government (the group) existed for the benefit of the individual.

"There are several theories as to why this was so, though the one I was most comfortable with was the influence of early capitalism. This was a system in which individual competition was praised above all else and winning was the important thing. The 19th century winners applied a distorted Darwinian explanation, "Survival of the Fittest." Further, the very important oppositional system to 'dog eat dog' capitalism which emerged in Europe was Marxism, in which theoretically the workers were going to take over. That this never happened is another story.

"Anyway the Americans took over the idea of the competitive capitalist from their Euro-brothers. This was perhaps best exemplified in literature by the American novelist, Ayn Rand, who wrote about powerful industrialists who let the chips fall where they would, which meant the weak would be left to fall by the wayside.

"Then when Americans started to expand over North America, they further glorified individualism by selecting the rugged, two fisted, fast-draw loner as their hero. He defeated all opposition alone, with his fists or guns. He was enshrined in Hollywood and continued as long as the American dream machine remained active. There were dozens of such heroes, though two of the best-known

201

were John Wayne and Clint Eastwood. And then there was the character of Rambo who defeated whole armies single-handedly."

"So rugged individualism was the American way?"

"I don't know that it was in actuality, Mar. Our so-called gunfighters were mostly ideals, creations of the entertainment industry that fulfilled a basic need. Historians said that the real fighters on the frontier shot their enemy in the back or with a shotgun, with which they couldn't miss. They did not face the enemy Eastwood-style, and then kill him with a fast draw. Hollywood films always had gunmen of all kinds shooting it out face to face."

"Okay Pete. So things outside Hollywood were changing rapidly toward the end of the century. And it seems you feel the Japanese were gaining dominance by their deep commitment to the group."

"That's what most people thought. And the evidence was impressive. For instance, the Americans had thought their dollar was the basis of international currency as had the British earlier of their pound. But at the end of the 20th century the Japanese yen and the German mark became co-equals. This was hard for the Americans to accept.

"However, the Americans became so concerned about the Japanese, they didn't see what was happening in the rest of East Asia. There were a number of mini-miracles in little countries with even fewer resources than the Japanese which began to produce like crazy, Taiwan, Hong Kong, Singapore, and South Korea. It got to the point, Mar, in the later 20th century when the 'Made in America' label was in the minority in American discount houses and specialty shops. These smaller industrial powers were called "the little dragons," presumably in comparison with the big dragon, Japan. But, of course, there was also 'the super dragon', mainland China. It got off to a slow start, but by the end of the century it was catching up in the production race.

"The producers of the world, including America, made the same error they had with Japan. They looked at the billion-plus population of China as a huge market where enormous amounts of goods could be sold. But the Chinese also had seen the writing on the wall, and as soon as they got tooled up, they started exporting. And who could compete with a billion hardworking, socially committed producer/consumers?

"By Takeover no one could be sure which would be the leader in production and finance, China, Japan, or some combination of the little dragons. But dominance certainly had shifted. The Euro-world was only one among several. This process was similar to the shift of power from the Mediterranean countries of Greece and Rome to Northern Europe when industrialization occurred.

"Yes Mar, no doubt about it. Most of the change had already taken place by the time the Atierrans arrived."

"I can almost guess what you believe was the prime cause, Pete. It must have been some version of group commitment as you described among the Japanese. Did the other East Asians operate in the same way?"

I recognized that she was fishing for the ultimate explanation; and she knew I would give it. I laughed, "Okay Mar, I will now put myself in the same boat as our intrepid leader at the end of W.W.II, General MacArthur. I thought about it a lot, just as he had first about the Japanese threat, then the little dragons, and finally the big dragon. I decided it must have been their social tendencies in general. You see, all of them tried to solve problems through group effort of one kind or another. Individualism was undesirable, if not dangerous.

"I can think of two good illustrations from Japan and China. The first was something my son said and did. He had studied hard to become a scholar of the Orient. But instead of going into academics, he opted to also study finance and follow the American dream. He became a banker and did well. But after a few years in his first job, he became discontented and took a job in a different bank. When he told me about it, I said, partly in jest, 'You know, old bean, you are violating a basic principle of the Japanese culture you studied. Job-jumping there is totally suspect. It indicates that the worker thinks more about his personal interests than the interests of the group.' "

His reply was, "I know, I know. But what can I do? I was brainwashed in this culture and I have its values."

He was right. He had read many of the same books I had. We had learned that there were two different ways to get ahead in work and other pursuits. There was a process called horizontal mobility in which one looked for another job if one did not like what one was doing, for whatever reason. That was the American way, whether in business, academia or marriage. A history of such changes did not hurt a person's record, even in marriage. It was okay to dump one wife and take another. The practice was based on individual choice and feelings. Social responsibility was hardly considered. Then there was the Japanese, or I would prefer, the Oriental way. It depended on a person staying with his company or marriage or whatever group through good times and bad. And the worker would be taken care of, be permitted to rise to his or her level of competence. And he or she would continue in his/her marriage. This process depended on commitment to the group. This gave a strength to Oriental companies that the Americans' lacked."

The more I thought of it, the more I realized I had been no different. Whenever I had been discontent with a job, I had looked for another. I also had been married three times. I realized it was one thing to recognize differences in another culture, quite another to change one's behavior accordingly.

"Good example, Pete. I can see how you must have got ideas across well when you were a professor. And you didn't shy away from personal reactions. That's good."

"Thanks Mar, but let me give you the other illustration. Its main value is that it shows that other Oriental societies also were committed to group rather than individual needs. This example had to do with getting rid of sparrows, believe it or not, and it was about the Chinese. I read about it in the New Yorker magazine first, but many years later I met an 'old China hand,' a retired Britisher who had lived in Hong Kong at the time and who still vividly remembered what had happened.

"The adaptation of the ordinary house sparrow was a great success story in the animal kingdom. Descended from the African weaver finch, it had traveled around the world on the ships of *sapiens*, mainly Euro I suspect, and learned to live off the droppings of its biped host. It left a bit of a mess with its droppings and woven grass nests in the crannies of buildings, and it also picked up grain or other edibles it found lying around. It was not popular with *sapiens*.

"The biped dominator never did care much for creatures which could take care of themselves. More often than not, if he didn't eat them, he treated them as pests. *Sapiens* really preferred those he could control. When he could bring them to the edge of extinction, he would categorize them as 'endangered species' and set up preserves to keep a few alive, and hopefully reproducing. Needless to say, the sparrow was considered a pest. It had few friends in the populations of *sapiens*, but it did okay on its own.

"Anyway, in the '60s the Chinese government decided to get rid of their sparrows. So they designated a day on which all citizens were supposed to kill them.. The technique was for each person to arm himself with a noisemaker, say a tin pan and spoon, and each time a sparrow was sighted trying to land, a din would be created. The idea was to exhaust the sparrows and then dispatch them. By the end of the day, all the sparrows would be destroyed.

"What intrigued me was that the Chinese would think of mobilizing their entire population to accomplish this task. It was an idea that was possible only among people who were very dedicated to social solutions. The Euros, who had little more love for the sparrow than did the Chinese, resorted to poison or traps or some other mechanical procedure devised by specialists. Euros did not usually resort to social solutions."

"Intriguing, Pete. But didn't you tell me that the Chinese got their current ideology, Marxism, from the Euros, or rather Russkies. How was that so social?"

"Well, Marxism was a social philosophy, but the Russkies, the original Marxists, certainly relied on mechanical and chemical solutions to problems when they could. But the Chinese had a double dose of social input, you might say. Before they adopted Marxism, they had followed the greatest social philosopher of all time, Confucius. And compounded with Marxism they had what could be called 'a double whammy.' "

"I see now, Pete. You are claiming that the Orientals were deep into social procedures and group commitment while the Euros were emphasizing individualism more, and that is what made the difference."

"That's it. I can't see it any other way. And that was the way the world was moving when your group came; it was in transition from West to East, from individual to group action."

15

The Last Chapter

I took in the outlines of the building as I approached the doorway. It wasn't anything to write home about, a rectangular structure about two stories high although I knew there was no upper floor. Since it had been used as a repertory theatre, it had needed height for the stage equipment. The walls had been stuccoed and a few windows put in at ground level, presumably since it had been taken over by the Atierrans. I was pleased to note a light coming from the recording studio. It probably meant that Mary was already on.

I went in and walked through to the couch which faced the screen. Mary was lit with the single world READY. I glanced at the corner to see if the red observation light was on. It was. Mary spoke right away, "Hi Pete. How do you like this, I got ready before you? This must be a first." This was followed by a slight giggle which surprised me until I remembered that the video-computer had simulated a female voice throughout, so why not a giggle?

I said, "Yup, things have changed. I suppose that your early readiness has to do with the fact that this is the last session. We'll be finished after today, right?"

"That's right, Pete. Even this one's extra, you know. It was made very clear to me by Central that after today the room will be used for another project. This is all the time we have." Pause. "Does that make you sad?"

I thought about it for a moment and then replied, "Yes, I think so. You know

206

I really got into our project." I studied Mary carefully and then went on, "And whether you know it or not, I've become rather attached to you, Mar. And you can believe that this is a first for me. I didn't grow up in the computer generation and never did get comfortable with one of these machines. So to get fond of one is wholly unexpected."

I swear there was a pink tint on the screen when Mary answered. I couldn't help thinking it was a computer blush. "Well thanks, Pete. I must say I appreciate that, and I must also say I have enjoyed working with you. I suppose your social scientists would say we really established good rapport, eh?"

I chuckled this time. Mary was using the jargon. However, I was becoming a little embarrassed. Trading compliments was not my style. So, "Okay Mar, today we will wrap it up. And I do think we will have done a pretty good history. But as always in an account or an ethnography or a history or just an occurrence, there are invariably some loose ends. So I feel fortunate that we have this chance to tie them up. And incidentally, this is your time especially, Mary. I know I have been hogging the show to a great extent, but now please ask me any questions that have occurred to you. That, incidentally, was the usual procedure with ethnographic studies, to keep going over the data as it accumulated and working up questions that occurred. And as you know, Mar, you've been a stand-in for the old boys of early ethnography who recorded the native cultures before the last old men died."

"Okay Pete, sounds good to me. Shall we proceed?"

"Sure. Fire away."

"A minor thing, but it still bothers me a little. In the standard terminology of your field, what have we been doing, a history or an ethnography?"

"I'll try to clear that up though as is not unusual among us long-time professors, I will start by claiming that it is more complex than it seems. We've certainly been doing a history, a history of the "white man's" impact on the "colored man's" cultures, with a little early history and prehistory thrown in to round out the story of *sapiens.* In the beginning you asked me to explain how *sapiens* achieved dominance over other species and how the particular kind of *sapiens* in control at the time of Takeover had achieved mastery over others.

"In actual fact, Mar, all existence is historical because things are always changing, and most simply put, that is what history is. Non-historical facts are figments of the mind. We describe relationships and things as if they had a permanence even while knowing that they are changing all the time, even while we are speaking. Non-historical facts could occur only if there was a state of timelessness, which we never see in the real world. The various afterworlds are something else but, of course, from the point of view of naturalism, they too are figments of the mind.

"But even so, we *sapiens* sometimes have found it useful to describe things as if they were timeless. We used to talk of the English language, for instance, as if it had a permanent structure even while we knew that it and all other languages were always changing. Or we talked about how the Japanese concentrated on group action while the Americans emphasized individualism. But we knew these too were in flux, that the degree of emphasis on individuals or groups was always changing in both cultures. Anyway, it has been found useful to describe cultures in this way for some purposes. In literature we used to call this the 'slice-of-life' approach, and in anthropology we called it ethnographic."

So anthropologists invented a time for describing native cultures which they called the ethnographic present. It was supposed to be 1492 when C. Columbus landed, after which, of course, the Euro-onslaught began and the major changes in native cultures occurred. Actually, the early anthropogs used this hypothetical time period simply as the time before gross interference by Euros began."

"And what about you, Pete? Have you been using this ethnographic approach?"

"Well yes, in a way. I've tried to describe the history of all the customs we've discussed, but I stopped at the new ethnographic present, 2020 A.D., when the Atierran conquest of earth was completed. The way customs were then would be my version of Euro-ethnography, the last ethnography, as I explained before."

PAUSE came onto Mary's screen and I assumed she was digesting what I'd said. I waited.

In a few moments Mary came on again. "Okay Pete, so we have been doing both a history and an ethnography. So on to the next. You had mentioned that your idea of what history was had changed. Is that right?"

"Yes, and that's something I want to get off my chest. Like so many others, Mar, I was raised with the idea that history was a true account of past events. It was only after I had studied anthropology and read histories written by 'others' that I realized most standard accounts were biased and ethnocentric, that they were generally versions of the past generated to justify the present. Even the verbal history of the tribals was like this. They spoke of mythical creatures of the past that produced all the good things of their earth, particularly themselves. These were like religious histories generally. Only a believer could find them reasonable.

"I must have started on the road of disbelief in the church of my boyhood. It became difficult for me to believe in the many strange events and 'miracles' the priests sermonized about. Could Jonah really have been swallowed by a whale and live? Could Noah have gathered all the animals of earth and put them on a primitive boat? Could Moses have brought plagues of snakes, frogs, and other vermin on the Pharaoh of Egypt by the flick of his wrist? Could he have parted the Red Sea and held back the waters while the Israelites crossed over? Could Jesus Christ have brought people back from the dead by laying on his hands? And could he have ascended into heaven by celestial levitation? I mean none of these things jived with the world I was seeing around me. And I just couldn't believe the stuff about miracles that the priest suggested. So I dropped out of the church."

"I must say I was still brainwashed enough to accept most secular history as really reflecting the past. Of course, I had little real history, biased or unbiased, under my belt at that time, especially of Asia, Africa, and Latin America. Mostly I had the popular history of high school."

"And then I drifted into anthropology where I heard a new version of history, especially of the tribal peoples. Each of the ethnographies I'd read was devoted primarily to a description of how life was in the old days, but at the end there was almost always a chapter about what had happened to the group in modern times. And almost invariably their culture had been destroyed or drastically altered, and the descendants relegated to destitution. This had happened to the pre-urban people all over the world, though being an American, I learned most about the Indians. Then when I saw the first Indian survivors, I knew that the mini-histories in the ethnographies were much more correct than what I had learned as a boy."

"In those standard histories I had been taught that the United States was a society which had been founded to promote liberty and freedom. I also had been told that this was a heritage from the ancient Greeks. At that time I did not question the fact that the American signers of the noble documents were almost all better off male landowners, many of whom were also slave-holders, and practically all of whom were Indian eliminators. Nor did it bother me when I first heard that the ruling class among the Greeks had been the same type, and that no woman ever spoke her piece in public, and that the majority of the population were slaves, and that Greece had extensive colonies of people they had conquered. Much later I learned that one of the most highly touted institutions the Euro-world had inherited from the ancient Greeks, the Olympic Games, had been quite savage in its treatment of women. Not only were none allowed to participate in or attend the games, but any woman caught peeking was thrown off a cliff."

"The usual version of Euro-American expansion was that North America was a vast and empty continent which needed filling up. Little did I know then that the whole continent was all ready occupied by Indians who when they realized the Euros were intent on taking over almost everything, resisted. The acts of resistance by Indians were most often referred to in Euro-history as 'Indian Massacres.' Little did I know that the most frequent 'massacre' was of the Indians, especially women and children, by Euro-militia or cavalry. The Indian men (braves) were harder to catch. But by killing women and children, the Indian problem could be solved efficiently. In between massacres, the Euros burned the cornfields of the Indians and killed the buffalo."

"Another idea was promulgated by Euro-Americans in regard to their expansionism, first in North America, then in Latin America, and then in the Pacific. It was referred to as manifest destiny. The term means 'clearly apparent' without specifying why and to whom. It enabled the Euro Americans to avoid explaining their various takeovers. What they did was simply self-evident. This was the kind of wordplay Euros generally indulged in to justify their actions. They practically never said what they really meant, that they had the power and they were going to use it."

Mary interrupted. "Okay Pete, I get the idea. The traditional version of history was usually quite different from what actually happened. But what has that to do with your version? Are you saying it is closer to the truth?"

I grinned self-consciously, "Okay Mar, you got me on that. But I'm going to stick to my guns. Without false modesty, I do think what I've related to you is nearer the truth when it comes to relations with the colored man. But I would prefer to consider it the anthropological view rather than mine even though I'm sure many anthropologists would disagree.

"You see, the anthropogs did not take the point of view of the dominants. By a fluke of their own history, they got started studying the losers, the tribals and civilizations that were temporarily weakened."

"But what has this to do with history, Pete?"

"Well, the real fact is that almost all histories have been written by the dominants, their versions justifying their actions. Most dominants were little interested in natives or what they called barbarians or savages. The only Greek who wrote enough about 'others' for anthropogs to quote was Herodotus. Non-Greeks were barbarians, especially the tribals. Or look at some of the Roman writings about their war against the Germanic barbarians. Who would have guessed that those savages were going to invent the scientific way and eventually conquer the earth? So the histories we know which describe the Euro-era have been written by the Euros for the most part, naturally presenting themselves in a good light. People who resisted their expansion were 'bad.' or at least wrong. In the folklore of the American frontier there were 'good' Indians and 'bad' Indians. Good ones did not resist the Euro-takeover, bad ones did. Of course, from the Indian point of view it would have been just the opposite. Not that there was much chance for success, but any Indians who willingly allowed Euros to take them over were cutting their own throats."

"Are you saying that this bias in reporting history was a particular characteristic of Americans?"

"No Mary, not at all. It was a general tendency. People of all cultures gave their children the versions that justified their own existence. As a matter of fact, the Americans were 'holier than thou' about the Russian version of history. I'm sure the Russians did rewrite history, but so did the Americans. The Americans did not burn books and took a lot of credit for not doing so, but they had other methods."

"So when you say that people changed history, you don't mean they took a given version and deliberately rewrote it."

"No, not always, although there were cultures which did burn books and then had new, approved versions written which eliminated undesirable incidents or persons, or changed them beyond recognition and replaced them with those they preferred. The Russkies' wartime prime minister, Josef Stalin, came into disfavor after his death and was at least down-played in later versions of Russian history. The same happened to the revolutionary leader of the Chinese, Mao Tse-tung. They even took down the public statues and photos of the leaders who were out of favor.

"The Western Euros had more subtle methods. Since histories were written consecutively, and the older versions discarded, the simplest thing to do to change them was to leave out or down-play occurrences that had gone out of favor and emphasize the newly approved ones. For example, when Americans came into the era of civil rights, they stopped writing about atrocities against Indians. I remember that in one of my earliest anthropology classes I was shocked when the old-time professor of Indian culture, Fay Cooper-Cole, claimed that massacres against Indians and other atrocities were fully documented in the old county records of my home state, Indiana. I had learned nothing about this in high school. Also, of course, historians could simply omit events that would be embarrassing.

"And one's glorious leaders became more glorious over time while the bad types got worse. The father of the country, Washington, couldn't tell a lie and the civil war president, Lincoln, walked a mile to pay back a penny.

"The actual basis for founding a society could be changed in later histories. Thus over time American society was presented as having been founded on liberty in general while in actual fact it had been founded on liberties for the privileged few well-to-do, white, Anglo-Saxon, heterosexual males. All the human rights that Americans espoused later for blacks, Indians, Latinos, Orientals, women, homosexuals were modifications of the original intentions, at least if we take the position of the founders into account.

"But Mar, let me get off the backs of the Americans. Others were no different. The great colonial powers such as the British, actually claimed that colonialism was for the benefit of the colonized. One of their favorite claims was that they had to take over these various cultures because the people were not yet ready to govern themselves. The colonizers didn't concern themselves with what the native cultures were doing before they showed up. They also did not talk much about the economic benefits being accrued by the British from the colonies. The British called one of the major uprisings in India the Sepoy Mutiny. When India became independent, it called the same conflict 'The First War of Independence.'

"Thus, when students from the colonized countries went to the home countries, they heard different versions from what they had heard at home. And though most accepted such, some mavericks tried to adjust what they had seen and learned in their own countries to what they were being taught in the centers of empire. And the fit was not good. A book which made a strong impression on me was The Discovery of India by Pandit Jawaharlal Nehru, the long-time prime minister of independent India. He wrote it in jail where the British put him for agitating for independence. And to make a long story short, he wrote about India's great achievements before the takeover by Britain, and about which he had heard little in his English education. One thousand years before the barbaric Anglo-Saxons invaded England, the Indians had developed a vast literature, a scientific study of language, at least eight systems of philosophy, two major religions, and artistic creations never equalled by the British.

"The Russkis also claimed that they were helping the central Asian nomads and others by taking over and passing on Marxism. Even the tribals had skewed versions of history which invariably proved that they were at least all right or even better then other tribals. Most hunting peoples' histories claimed that it was their destiny (manifest?) to kill and eat the animals around them. This doesn't sound too different from what the Judeo-Christian-Muslims taught their people about meat animals in their holy books."

"So there we are, eh Pete, carrying on with biased versions of history? How do we know what the truth was?"

"Well Mary, I think that the only thing *sapiens* could have hoped for would have been reduction of ethnocentrism and a willingness to see that even the dominants had some imperfections. And this, of course, means the people who were writing the histories.

"I know it would have been asking a great deal of Americans to accept President Andrew Jackson as primarily a paramount Indian basher, and Abraham Lincoln as an astute politician rather than a Horatio Alger from the frontier who became 'the great emancipator.' The slaves were freed by Lincoln primarily out of political expediency, you know. When he issued the Proclamation, the North did not even control the South. So the piece of paper was just that. Freedom for blacks was to be a long, drawn-out process. And the idea of Lincoln being a roughhewn 'rail splitter' from the backwoods was an electioneering gimmick. Anyway, he wasn't too bad as a president, though his policies got many innocent men killed. However, that had always been the lot of the common man who served in national armies. But Lincoln's image certainly changed in later historical versions."

"But how does one decrease this ethnocentrism, this tendency to use history as a means of justifying the present?"

"That's one I thought about a lot, Mary. One of the main problems, it seems to me, is rooted in morality. The Euros, especially the Americans, saw themselves as a very righteous people, as nice guys. They kept telling each other that they were always doing 'good' things as a matter of national policy. And that took some very fancy dancing when you consider their expansionistic tendencies. It was impossible to take over another people, dispossess and deculture them, make them work for you or buy your cheaply made goods, and still be moralistic nice guys. But if on the other hand, you admitted that power had its own rules just as in biological natural selection, and that morality had little to do with national policies, you no longer had an inconsistency. Actually, some writers had said as much in the past, but they were quickly hushed up or discredited. Probably the best known was Machiavelli, an Italian who wrote about the techniques for gaining and using power. He did not draw on moral principles, and stirred up the other Euros so much, they coined a word to signify bad actions, 'Machiavellianism'."

"Now I admit this was a bitter pill for the Euros to swallow, but were they to live to the end of their dominance in an illusory world? Power was the source of control and I don't mean moral power, I mean military, technical, social, and commercial power."

"They could have tried to reduce the degree of ethnocentrism that went into the average citizen's brainwashing. Now, while I know that the members of a culture had to have some pride in their system so they would struggle to continue it, too much self-centeredness turned out to be a negative in world history. It drove members to insist on their own way no matter what. And this brought on violent cultural excesses such as mass slaughter of 'others,' extensive wars, and great exploitation.

"Also, the fact was that dominance was always temporary, just as it was in the biological world. Breeding of social mammals was always controlled by the male who was physically the most fit. However, he had only a few years of dominance, inevitably being replaced by a newly matured male who had become stronger. And in social groups, dominance inevitably passed from one to another as the balance of cultural power changed. Civilizations came, stayed a few hundred years, then died out, to be replaced by new ones. This process inevitably was resisted by the losers, as it was in a biological group, but to no avail. Among *sapiens* one of the reasons the weaker cultures resisted takeovers so strongly was that they invariably thought of their culture as superior, no matter how high or low it actually was. Thus the smallest tribe resisted as much as did the largest civilization, at least at first.'"

I paused, but then added, "And now the Atierrans. But, of course, the power difference between them and the cultures of *sapiens* was so great from the beginning that ethnocentric views of superiority were out of the question."

We both stopped at the same time. On Mary's screen was the word PAUSE which I didn't fully understand. Had she decided there was nothing more to say? There were still two more issues I wanted to cover, so I said, "Mary, I do want to put on record the role of my own profession because I think it was significant."

"Okay Pete, that's all right. We still have a couple of hours. Go ahead."

"Well, first I must say that it was almost inevitable that an anthropologist would be the first to try to present this kind of history. Those of almost all other professions were so caught up in the values of their own cultures, they could not see history from the other side. They had been too well brainwashed."

"And anthropogs? I hope you don't mind my using the term, Pete."

"No, that's okay, Mary. I've told you I didn't take my profession that seriously even though I think it had some important information to contribute. But what high muck-a-muck in the State Department would listen to an anthropology?" I sighed. "Anyway, that's the way it was."

"But to get back to what the anthropogs had to offer, I've already told you that they studied 'the others,' mainly tribal people at first and later complex cultures that were temporarily in disarray. Besides the tribals, the main other societies studied were the colonized. While the sociologists, our nearest professional relatives, were studying their own people, we were studying people of other cultures. And to get anywhere in academia and a few other places, we had to justify what we were doing. So we came out with an idea we called 'cultural relativism' which simply meant that each culture had to be judged according to its own values. Anthropogs actually came to justify cannibalism, vegetarianism, primitive warfare, bizarre sexual practices, and no end of supernatural beliefs and practices. It became one of the anthropology's roles to justify the ways of others when they were considerably different from the ways of Euros. Some anthropogs, made their reputation by proving the logic of the most outlandish customs. In fact, many anthropogs came close to reverse ethnocentrism in claiming that many customs of primitive societies were superior to those of civilized people. Such arguments rarely asked why the primitive societies were relatively powerless. And though I believe anthropogs sometimes went too far in their defense of 'native' customs, they were the nearest to being non-ethnocentric in a world where ethnocentrism was the norm."

"So most ethnographers wrote mini-histories of 'their people,' including the time they came into contact with the Euros, and basically from the point of view of the natives. All I've done in this report is to take a world view of the Euro-influence by using my knowledge of these mini-histories."

"So anthropology was the least ethnocentric of the social sciences? And you think for that reason alone it had something to offer."

"Well, it might have, Mar, if the matter of survival had not been taken out of control of *sapiens* by the Atierrans."

"But even so, anthropology was a kind of mixed bag, as we used to say. No matter that we may have taken the native point of view more often; the fact was that our studies were the consequence of our being Euros. Until the middle of the 20th century it was rare indeed for non-Euro anthropogs, and there were a few, to study any culture but their own or their own tribals."

"The study of other people became fashionable only in the 19th century after Euros were in control throughout the world. So the curious could go to all the outback places they wanted and watch natives and ask questions. And whether they admitted it or not, the anthropogs did their field studies under the umbrella of the Euro-gun. Natives had learned by then that it was risky to oppose the light-skinned, hairy bipeds, even if they were inquisitive investigators who claimed they favored natives. Basically their main form of resistance was to refuse to answer questions or to lie, the ultimate defense of the weak. Even so, because there were always some natives who would cooperate, the stack of ethnographies grew higher."

"So the bread and butter of anthropology became the ethnographic report, accumulated from interviews and observations of native customs. And how anthropogs did professionally was in direct proportion to the acceptance of these reports by their fellows."

"Moreover, it was probably inevitable that anthropogs would become the spokespersons for natives. After all, in the beginning their livelihood depended totally on the existence of tribals who could be studied."

"The usual attitude of Euros during most of their history with the colored man was to subdue him, divest him of whatever wealth he had, mainly land, capture his soul, or use him as a worker or customer. And then along came this Euro-guy in a pith helmet or Stetson hat, carrying a notebook and camera, and asking questions. What a difference!"

I waited to see if Mary would respond. When she did not, I went on, "Well, actually the anthropogs were like other Euros in one respect. They were acting on the basis of self-interest. They was getting their credentials, recognition by their fellows, and a stable position in a university."

"Are you saying then, Pete, that the anthropogs were completely self-serving in their studies?"

"No Mar, I don't believe that. But then I don't believe the other Euros were either. Now, while it can be said that the colonial Britisher couldn't have had much concern for the Chinese to have pushed opium onto them, his mercantile brother who was pushing tea onto the East Indians probably thought it would benefit them. After all, the British liked tea. And though the American slaveholder who broke up families couldn't have been much concerned with the happiness of his blacks in this regard, he might still have felt righteous about helping them get the old time religion. And even the missionary intent on capturing a place for himself 'on the right hand of God' or in that vicinity, probably still believed he was helping the nature worshipping tribal by bringing him Christianity. So the anthropogs who were gathering material for their Ph.D. dissertations might still have believed that publication ultimately would help the natives. And some anthropogs went further in their concern. They protested bad treatment of natives. North American anthropogs agitated about the atrocities committed on the few remaining tribals in Brazil by the settlers (land-grabbers) and modernizers (road builders) in the latter half of the 20th century. Of course the anthropogs were sounding off from the United States where the land had long been taken from the natives and thoroughly ravaged. But still, apart from their need to study the few remaining tribals in Brazil, I suspect that most anthropogs who spoke out had some genuine feeling for the Indians. Whether they would have had as much if their tenured professorships had been at stake is another question."

"One fair-sized group of North American anthropogs became expert witnesses for the natives in legal battles to get restitution for land loss. And this also took place in the latter half of the 20th century, more than 70 years after 95% of the Indian land had been expropriated. Euro-Americans were then having some guilt pangs for the hardships they had caused most non-Euros. A law was passed to permit some redress for treaty violations and other crude land-grabbing procedures of the 19th century. Some Indian groups did get some money as a result, but none got any appreciable amount of land."

"It seems that the anthropogs were doing more for native peoples than any other professionals," Mary interjected.

"I think so and even while freely admitting they had a vested interest in taking the side of the tribals. For instance, if they had not been so busy writing the tribal ethnographies, knowledge of their way of life before the arrival of Euroman would have been lost forever. The Indians were not writing about their own cultures; they were too miserable to want to preserve any knowledge of the past, even those who had the ability to do so. So, by the mid-20th century if you wanted to find out how things had been, you had to go to an ethnography done by an anthropologist."

"Sounds like the tribals owed the anthropogs a lot, Pete. Were they appreciative?"

"Actually many were not. For one thing, the ex-tribals were just trying to stay afloat in Euro-man's world. It was clear to them that the old way of life was never going to return. Euros in the cities far from Indian country had guilt feelings about the treatment of the redmen, but Euros who lived where there were fair numbers of Indians rarely wasted such feelings. The old adage 'the only good Indian is a dead Indian' had been changed to 'the only good Indian is somewhere else!'"

"So later in the 20th century, some Indians got irritated with anthropogs hanging around watching, asking questions and taking pictures, then returning to their universities to write more ethnographies. I once heard a definition of a Navaho family: a husband, a wife, a couple of children, and an anthropologist under the bed."

"The Indians complained that the only thing these people were interested in were native customs when the real need of the natives was to adjust to Euro-society by learning Euro-customs. They also complained that most anthropogs did little to help Indians on reservations. And though it was probably hard for an Indian to like a Euro-landgrabber (settler), at least he was not a hypocrite about what he was doing."

"So now Pete, you are giving me the dark side of your profession?"

"Well, you know I was selected for this job and promised to tell it the way it was. And you know my feelings about historical accounts generally that they are sanitized at best and falsified at worst. At the very least they practically always describe the people of the dominant culture as nice guys, never as self-serving exploiters. And I must say I have not seen many 'nice guys' when it came to dealing with natives people.' I'm afraid that Christian morality notwithstanding, self-interest was the primary moving force. Thus, anthropogs were no worse than others, and perhaps even a little better. But they certainly made good targets for the newly literate natives."

"I know from firsthand experience the ill feelings of Indians. I was thrown off the first reservation I went to as an anthropology student. The Ute Indians of southwestern Colorado had enough by the time I showed up with my professor. He was planning to take measurements and blood samples for a study of their racial characteristics; I was to analyze their language. Quite simply they voted us off the reservation, claiming that my professor was a German spy. W.W.II had just ended and my professor, having grown up in Germany, still had an accent. However, my guess is that they would have found any excuse. What Ute would have wanted to give a blood sample for the benefit of a professor from a midwestern university? Or for that matter, why should they have helped a college student learn their language? Who could guess what nefarious use he would make of it? Anyway, that was the unpromising beginning of my first field study."

"You have really gone on long about the Indians, Pete. And while I know you and other Euro-American anthropogs studied them most, this study, this last ethnography if you will, is about mankind in general. What about some of the other natives?"

"Good question, Mar. And I think it can be answered fairly quickly. In general, they acted much like the North American Indians and for most of the same reasons. Anthropogs were becoming well known from their work in describing their old way of life while the ex-tribals, poor souls, were struggling to keep afloat in the new Euro-world. In many places, and particularly in Africa, anthropogs wore out their welcome after the end of colonialism. The native leaders looked to sociologists, economists, and other social scientists to help them because they were more interested in modernization than in old customs. I know that to be true, also from personal experience, this time on a project to Nigeria under the umbrella of rural sociologists, communication specialists, statisticians, economists, and other modernizing types. The Nigerian leaders wanted nothing to do with anthropologists and I kept my background in low relief. Anthropologists got the same reputation in the newly reviving civilizations also, that they were only suitable for studying primitive tribes. Countries of the Far East, Near East, and even Latin America developed their own sociologists to study civilized life while their anthropologists studied what remained of their tribals. In the case of Latin America, these tribals were not very numerous because the Hispanics had applied the final solution to quite a number of them."

"In a way it is surprising that anthropology retained the reputation it did. After all, the request for an anthropologist as the unifying social scientist for an analysis of the rise to dominance of *sapiens* and Euroman came from the central bureau of Atierran intelligence. Somebody up there had gotten the idea that the best account of mankind's history would come from anthropology."

"Whoever it was, I think was right. For the broad picture of mankind in both time and space one had to go to anthropology. The other social sciences were much more specific as well as more ethnocentric, even if they generally did know more about civilized life."

We both fell silent then, more or less at the end of the topic, I figured. Since Mary's screen read PAUSE for quite a while, I lit my pipe. There was still one part of the story I felt was unfinished, perhaps because of my years as an applied anthropologist. I believed that at the end of a report, I should make some policy recommendations. Fortunately, Mary was enough in tune with my thoughts by this time and was thinking along the same lines. She said, "Pete, I'm interested in one more thing. What do you think would have happened and what would you have recommended to people of your own culture if we Atierrans had not taken over. What would you have done about the transition to come?"

I chuckled, "Ah yes, the quest to maintain dominance. It was an important issue, Mar, and one which I and many others thought about a lot. So first off, let me say that as far as I know, no culture ever willingly gave up its position of dominance. The Euros looked for any means to resist the transition to the Far East. But solutions were not easy. For one thing, a culture tends to have a central focus which is difficult to change. We in anthropology used to call them configurations or values. Perhaps our most literate analyst of the subject was a professor named Ruth Benedict. I went to Columbia University to study under her after reading her famous book, Patterns of Culture, but unfortunately she died before I got there to begin my studies. I still remember her writing though, despite the decades of Benedict-bashing that followed. There is no doubt that she exaggerated some of her theses and neglected to report some information, but I still think her approach was basically valid. She did a book about Japan that was particularly vivid, The Chrysanthemum and the Sword."

"Anyway, a given culture would evolve a certain tendency which permeated most aspects of life. And this became so deeply ingrained, it was very difficult to change. I suspect that the focus on individuality in Euro-culture would have been just as difficult to change as the focus on group responsibility in the Far East. Neither could easily take on the value of the other."

"And there was no inclination by the Far Easterners to take on Euro-individualism since they were succeeding. And though the Euros, particularly the

Americans, tried to take on some customs of the Japanese, they never even considered changing their basic value on individuality."

"Also at about this time America evolved a system of individualistic power grabbing, called hostile takeovers, in this case of corporations, which was thoroughly unproductive but made a few individuals quite rich."

"So you're saying, Pete, that fundamental change of the basic pattern was difficult?"

"I think so. After all, the Euros had been so successful for 500 years, how could they conceive of having to make fundamental changes? Dominants ordinarily expected others to change."

"So you are saying then, that the outcome was inevitable, that the shift to power had to occur?"

"Okay Mary, now we are getting to the nitty-gritty. Let me tell you first that I heard many predictions and recommendations in my lifetime that turned out to be wrong. Thus, I do not believe that *sapiens* had any great skill in this regard. As I have related, American policy toward Japan was basically wrong for 150 years in the sense that the outcomes were invariably different from what were expected. The same could be said of many other Euro-policies. The looting of the great civilizations of the Americas by Spain certainly did not make the looter rich in the long run. Euro-slavery was a policy that did not pay off for long, and the whole colonial gambit turned out to be a losing proposition. The great colonizer, England, ultimately was dragged down by its possessions.

"But even so, like others, I'll stick my neck out. After all, I wasn't an applied anthropologist for nothing all those years."

"Okay Pete, shoot."

"First off, we could have refused to modify our behavior. We could have insisted on continuing to be dominant, maintaining the same values and procedures that had made us great. But that would have been unlikely since enough people in the Euro-world could see that something different had to be done."

"Or we could have taken direct action, to resist in the one way still available, military action. Euro-Americans and Russkis still had the most humungous war machines. And the Euros had always relied ultimately on military solutions to international problems, twice with Japan. The spirits of Commodore Perry and General MacArthur gave testimony to that. And I think this might have been a possibility a third time. A good proportion of the Euro-population, and again specifically the Americans, were into verbal Japan-bashing. They told each other that their predicament was the fault of the Japanese, despite all the good things Americans had done for the sons of Nippon. Americans were particularly incensed that their country had invested heavily to defend Japan and kept the sea lanes open, especially to deliver the oil they needed. Japan had no army or navy, but most such critics conveniently forgot that rebuilding their military machine had been expressly forbidden 'forever' by the constitution that General MacArthur forced the Japanese to sign. Anyway, I doubt that this scenario would have worked, mainly because the weaponry of that period was so destructive. After the Americans had devastated two Japanese cities, nobody was willing to start a nuclear war. No major wars had been fought since, largely I believe out of fear of a nuclear conflagration."

"Okay Pete, what then? You've already indicated that the Euros as dominants were unlikely to simply accept the new order."

"The only instances I ever heard of when one culture willingly gave up its own customs to adopt the ways of another was when the incoming power was enormously greater. I believe I mentioned to you that the Cherokee Indians of the American Southeast went through a voluntary transition to Euro-American culture, not that it helped them in the least. The Euros displaced them anyway. However, most Indian groups did all they could to maintain their own way."

I looked straight at Mary's video, knowing that my face was being photographed all the time. I said, "And of course, we *sapiens* are now facing that kind of power from the Atierrans. And since it is impossible to resist, I suspect we will go through a transition, that is, if the Atierrans let us." I knew that many tribal groups had not been given that chance by the Euros. They had been wiped out to the last person."

"Anyway, the threat from the Far East wasn't as great as that from the Atierrans. There even seemed some possibility that the Euros could maintain their independence in some state of co-dominance. I would have believed that to be the most desirable future for Euros, the most they could hope for."

"That doesn't sound too bad even if it would be a considerable comedown," Mary commented.

"I don't think it would be either. I think the exclusive dominance of the Euros for almost 500 years was quite a respectable chunk of history. The Euros could have tried to merge, to create a blend from a combination of cultures. Although there had been many other instances. I'm thinking of Graeco-Roman civilization in particular. In that case the Greeks lost the military war and were taken over administratively by the Romans. However, the Romans took over many aspects of Greek culture, which survived even after Roman domination had ended. I suppose as a parallel, if dominance had gone to the Far Easterners in the war, they might have tried to adopt even more Euro-science, medicine, architecture, engineering, and even more music and the arts. It was fairly certain anyway that they would have continued with the main Euro-language, English. And what would the Euros have had to do in return? Remember, by then the Euros were no longer in a position to demand anything. So, above all they would have had to cut back on their individualistic, confrontational, non-compromising, nonsocial way of solving problems, and try group efforts in meaningful ways. I know this would have been a great deal to ask. The whole Euro-legal system would have had to be overhauled to emphasize compromise instead of confrontation, the workers would have had to cooperate with managers, trade unions would have had to be changed or eliminated, the hostile takeover would have had to be abandoned in favor of policies that would improve production, the hostile divorce would have had to be given up in favor of the family compromise, and children would have had to be indoctrinated to believe that education was the highest good in the early years, a contributory force to the success of the family and society. In short, individuals would have had to be concerned with the social unit rather than the self. Above all else, one could not concentrate on 'doing one's own thing.' The values of Henry David Thoreau, Ayn Rand, and the frontier gunfighter would have had to be abandoned."

"Wow Pete, that sounds like a lot. And would your standard Euro have been capable of so much change?"

I sighed. "I'm not sure, Mar. I know it's a lot and I also know the tremendous force of cultural inertia on the average person who normally tried to keep doing what he had been brainwashed to believe was proper. And since most Euro-institutions were based on individual rights, basic institutional change also would have been necessary. On the other hand, I doubt that the East Asians would have become self-seeking individualists. They too were constrained to follow their own cultural values, which unfortunately for the Euro-world at that time were more successful." I paused. "But what else was open?"

"Okay, I understand Pete. And so with that, your final pronouncement, I take it the case of the Euros is closed."

"Yes, I have nothing more to say." But then I just had to put in a plug. "One last thing though, Mar. As you know, we have done this account primarily because I am an anthropologist. From this viewpoint it was possible to look at the 'others,' the colored men of the world, with a minimum of ethnocentrism, to see their customs and reactions to the Euro-onslaught as rational procedures. For the first time in world history we had a fullblown study of mankind which looked outward rather than inward. The 'others' were not judged by the values of the dominants. And for the first time we could see the 'others' without great prejudice. And among other things, we could understand them better. I would venture to say that if anthropogs or other non-ethnocentrists had been called in to help make policy decisions, there would have been fewer errors in evaluation of the Japanese and other Far Easterners. And to add to this, I have to say that if I have any noteworthy qualities, including the capacity to make this document, it is only because I followed the path of anthropology."

"So as long as the Euro-colleges and even high schools had continued, I would have required all students to take some anthropology or other non-ethnocentric courses. And if Euros had been given the opportunity to make any further policy decisions in the 21st century, they might have been less ethnocentric. The way of 'others' would at least have been seriously considered."

And then we both stopped simultaneously, almost as if at a signal. I thought it now was truly over and was wondering what to do next. Was I to say, "Thank you, it's been nice knowing you. Goodbye?" And was I then to get up, walk out of the room of the last ethnography and return to my little apartment, and to what? Mary broke the impasse as she had done so often before. "Well, I guess that's it, Pete. And I must say I enjoyed it." She giggled, "You see, we modern computers do have feelings."

"Okay, and thank you, Mary. I also enjoyed it."

I sat there a little longer, but she said nothing further, so I got up. Then she spoke, "So now, Pete, what's on your mind? What do you plan to do?"

I turned and gazed at the words on her video. "Good question, Mar. You know I'm pushing 100 and I probably have only a few more years to go. And also having been a cross-culturalist for quite a while, I have fretted less than others about the new takeover. I even could have lived with the Far Easterners if they had become dominant. Even my avocado farming would have fit in. The Japanese were crazy about this fruit and by the end of the century were no longer interested in farming themselves. I wouldn't have been unhappy growing avocados for their market. I think I could even have worked out a *modus vivendi* with the Atierrans, despite the bizarre nature of some of their characteristics.

"And so for whatever it is worth, probably not much, I think I'll go back to my intellectual roots. If there's no objection, I shall take a look at Atierran culture. After all, that's about all I'm good for, snooping into other peoples' business and then doing reports. I think first I'll go through the video-cassette collection and view all the visuals I can find on Atierran doings. Then if I can wangle it, I'll try to get an informant and start in on the language. That's the first project I undertook at the beginning of my linguistics/anthropology career, a study of spoken Norwegian.

"You know it is still bizarre to me to be surrounded by creatures who communicate very efficiently with a complete absence of sound. The vibrational communication of the Atierrans is totally unlike any ever reported on earth. I am fascinated by the continuously waving cilia which remind me so much of my sea anemone picking up food particles. But the idea that they are picking up vibrations of communication! And then I still don't know how the vibrations are produced, presumably from an internal or invisible organ. And there are so many other questions. Do the Atierrans communicate with ordinary symbols or do they have super symbols of some sort? Can they deliberately project emotions if they have them? And is there some way they can screen out 'noise.' And how far can they project these vibrations? Is there some kind of amplification device?"

Then I stopped, realizing I was getting over-enthused again.

"Okay Pete. Sounds interesting and best of luck on your next project."

TAKEOVER

INDEX

Order Form

Telephone Orders:
> 800-616-7457. Have your Visa or MasterCard ready.

Fax Orders:
> 619-728-6002

Postal Orders:
> The Hominid Press
> Arthur Niehoff
> Box 1481-16
> Bonsall, CA 92003-1481, USA

Please send the following books. I understand that I may return any for a full refund – no questions asked.

_____

_____

_____

Teachers can get an examination copy for 30 days. Please call or write.

School/Company Name: _____

Name: _____

Address: _____

City / State / Zip: _____

State Tax:
> Please add 7.75% (1.00 per book) for books to California addresses.

Shipping:
> $2.00 for the first book and 75 cents for each additional one.
> (It may take three to four weeks.)
> Air Mail: $3.50 per book.

Payment:
> ❑ Check
> ❑ Credit Card: ❑ Visa ❑ MasterCard
>
> Card number: _____
>
> Name on Card: _____
>
> Expiration date: _____